The Technological Conscience

The Technological Conscience
Survival and Dignity in an Age of Expertise

Manfred Stanley

The University of Chicago Press
Chicago and London

This Phoenix edition published by arrangement with The Free Press,
a Division of Macmillan Publishing Co., Inc.

The University of Chicago Press, Chicago 60637
The University of Chicago Press, Ltd., London

87 86 85 84 83 82 81   1 2 3 4 5

Library of Congress Cataloging in Publication Data

Stanley, Manfred.
   The technological conscience.

   Bibliography: p. 255.
   Includes index.
   1. Technology—Social aspects. I. Title.
T14.5.S7   1981   303.4'83   81–8199
ISBN 0–226–77096–6 (pbk.)   AACR2

Copyright Acknowledgments

82035

This book is dedicated to the memory of my mother, Ilse Stanley. In the concrete example of her heroic existence, I learned what it means to prevail as a person against all odds.

Contents

Preface

Another book on the fear of technology? What, one might ask, is there left to say? Yet, despite the plethora of volumes on the subject over generations—indeed, over centuries, if one reckons back to Plato's writings against the Sophists—it is the author's conviction that much remains to be said.

Broadly speaking, most recent books directed against technology have fallen into three categories. Books in the pastoral style evoke bucolic or primitivist fantasies against the sinister majesty of a technocratic imperium consolidating around us. In works that we might call pessimistic "technology assessments," we are warned that some modes of technology (e.g., computers, automated factories, drugs, television, transportation, weaponry) will so rearrange the landscape of mind and matter that, whatever we may now desire, we are all fated to adapt our behavioral rhythms to the harmonies these impose. Finally there is the school of humanistic pessimism, in which a doomed tradition of sacred or secular classical values is counterpoised against an emerging "wasteland" of "instrumental reason."

Against these pessimistic images, there are balancing optimistic perspectives. The pastoral critique is negated by celebrations of "the secular city." The pessimistic technology assessors are reassured by their optimistic counterparts. And the Cassandras of humanism are scorned as gnostic mystifiers by the grand inquisitors of "scientific" collectivism.

This great struggle of prophecy regarding the end of what all seem implicitly to agree is an era of "transition" cuts across the usual boundaries of political and cultural ideology. Pessimists and optimists about technology are to be found among partisans of "left" and "right," of "high" and "low" culture, of "collectivism" and "individualism," and of "sacred" and "secular" churches.

This book arises out of the conviction that a literature is now in order eschewing apocalyptic frenzies of doom or salvation in favor of calmer analysis.[1] It should not be inferred from this remark, however, that the present book is altogether "value-free." For one thing, although its author is not a "pessimist" and the book is hostile alike to pessimist and optimist fervor, it is most improbable that anyone who is blandly optimistic about (or indifferent to) the fate of the human soul in technological civilization will read this book with much patience. I frankly do not regard it as worthwhile to spend much time intellectualizing about why some people are sanguine about the future. The pessimistic case seems to me on the face of it far more informed by humanistic sensitivities, and no defense of this view will be made here.

But the pessimists seem to need rescue from some of their own blind alleys. If human beings are free, how can they be so enslaved or conditioned by technology as to warrant the near total rejection of modern trends that we find among many pessimists? If humans are subject to such powerful forces of determinism that a technocratic imperium is actually probable, then maybe the pessimists are pessimistic because they secretly fear that their "scientistic" enemies are more right about human nature than they are. At any rate, it seems that the pessimist case needs to be interpreted by a sympathetic listener who can avoid the very traps that humanists like to ascribe to their "technocratic" opponents. To provide such an interpretation is one of the purposes of this book.

The term "technicism," although in increasing use, has yet to be embraced by a single canonical definition. Skeptics can hardly be blamed for suspecting the term to be a euphemism for whatever the speaker happens not to admire that day about modern civilization. Anyone who presumes to write about technicism should feel obliged to venture a definition of what he assumes he is writing about. Providing such a definition is another purpose of this book, and the task is undertaken with the hope of rescuing the terms "science" and "technology" from the hostility directed by humanists against the phenomenon of technicism.

A book about a subject as vast as technology and technicism must, if it is to be reasonably well grounded, make certain choices at the outset. Thus, any aspect of the topic of technology can be addressed in the context of culture or that of society. In the former case one is interested in such matters as language and symbolism, styles, norms, values, modes of artistic expression, and so forth; in the latter case one is concerned with roles, organizations, and structures of rank and specialization. Another distinction in terms of which technology can be addressed is that between tangible and intangible modes. In the first case one is interested in physical technologies (transport, factories, communications, etc.). In the second case the focus is on nonphysical features of technologies such as technical languages, reality simulation rules, and bureaucratic organization maps. Finally, technology can be addressed in terms of its significance either for the fate of the individual person or with respect to some important aspect of the community, society, or civilization. These distinctions help fix the restrictive commitments of this book.

This book is about the culture, not the social organization, of technicism; it focuses on intangible features of technicism; and it addresses itself to personal fate, personal destiny, in a technicist environment. More specifically, it is about *linguistic* technicism—the misuse of scientific and technological vocabularies with regard to human activities better described in other ways. This focus reflects my view that the essence of technicism is not some fateful technological determinism but rather a mistaken understanding of the uses of language, arising not out of willful deception or ignorance but as a result of the slower evolution of the human understanding of language as compared with other kinds of evolution (e.g., technological, scientific, and social). It is a peculiarity of this century's intellectual history that we have become more sensitive than at perhaps any other period to language itself as a form of community, as a form of consciousness (and false consciousness), and as an instrument of world creation, destruction, corruption, and control. I say this in full awareness that in one sense we have returned to a primordial insight: primitive tribe and ancient civilization alike were characterized by respect for the magical powers of the Word. But faith in magic is not the same thing as secular respect based on theoretical understanding. I believe that the evolution of creative and responsible power over language among democratic polities is the next great challenge of moral progress. If this challenge is not met, democracy itself will succumb to subversion through linguistic self-mystification.

This book, then, is based on the assumption that technicism consists in metaphorical misapplications of some of the assumptions, imagery, and linguistic habits of science and technology to areas of discourse in which such mistakes obscure the free and responsible nature of human action. As such, technicism is a break in the evolution of linguistic understanding and self-control, a cul-de-sac of mystification. With this book, I hope to make a small contribution to the rational evolution of human self-direction.

Even given this restriction, the book could have been about many different things. It could have been about standards of expertise, the grounds of professional legitimacy, or the fate of nontechnological claims to authority.[2] But what this book is primarily about is the fate of the individual person: the powers of the human being, the dignity of the person, and the possible technicist mystification of personal consciousness under conditions of modern industrial civilization. If this restriction is not kept in mind, the thread that ties many disparate discussions in this book together will be lost to the reader.

The book is written in three parts. Part I defines the problem of technicism by setting it in a variety of contexts. In Chapter 1, I locate my understanding of technicism as a cultural phenomenon within a matrix of other plausible definitions. In Chapter 2, I venture an interpretation of the historical evolution of technicism, an evolutionary account that seems to me implicit in the fears and diagnoses of the pessimists. This account reflects two points I want to clarify about the pessimists' case: first, their fears reflect a coherent diagnosis, not just moralistic responses to current social problems; and second, the tendency of many pessimists to attack sociology as part of the technicist mentality rather than a response to it needs to be understood by sociologists as something more than simply an antiscience reflex. It should be seen, rather, as an invitation to sociologists to reflect on the moral ends, functions, and limits of sociology as a historically evolving discipline. Finally, in Chapter 3, I set the problem of technicism in its moral-philosophical context. It is important to recognize that technicism is based on a moral philosophy; it is not simply a philistine manifestation of narrow social interests. The interesting forms of technicism reflect a cultural revival of the natural law tradition, the effort to derive moral conclusions from the study of nature. Technicism reflects a particular attitude toward the proper relationship between science and conscience, an attitude that stretches in modern sociology at least from Henri de Saint-Simon to Emile Durkheim. In Chapter 3, I set forth the moral-philosophical challenge that technicism poses to the humanist imagination: how can the values of collective survival and personal dignity be reconciled? The problem of survival has received much attention in the general literature; the notion of dignity, little more than rhetorical salutes. Any concern for the fate of the person in a complex technical civilization requires a more than merely sentimental reference to what the claims of dignity are (and are not) about. In this chapter I try to demonstrate some important connections between dignity and the moral significance of persons as language users. Without this foundation, no consistent moral critique of technicism seems possible to me. Expertise as such must then be either blindly accepted or equally blindly rejected.

In Part II I turn to the question of what the arguments of Part I signify for the language of sociology. Again, these chapters could have been written differently, depending upon the central focus of the book. My focus being the dignity of the person, the chapters in Part II examine sociological language from this metatheoretical standpoint. In Chapters 4 and 5 I examine what a personal agent-centered sociological perspective is about. In doing so I try to avoid confusing personalistic sociology with a reductionist psychological approach to the field. For reasons cited in the text, this intention is best achieved by an intensive examination of the concepts of legitimation and scarcity. In this context, the general problem of expertise and its legitimation is investigated. Chapter 6 is a detailed case study of the intellectual uses and misuses of cybernetics as a metaphor of order. The point of the chapter is to document the sort of metaphorical misuses that are the source of technicist mystification. I seek to demonstrate how the person-as-agent can be made to disappear almost entirely from theoretical attention when moral preoccupation with collective survival and social engineering induces the humanistic disciplines to be unduly influenced by models drawn from physical sciences.

Part III is concerned with the question of countertechnicist intellectual policy and practice. Education, conceived as a panacea for all problems, has a virtually sacred status in American culture. Yet education has obvious importance for the transmission of cultural memory and cognitive habits of language use. This makes impossible a totally cynical response to the American fascination with education. If there is any hope for a cultural antitechnicist reaction, education, for better or for worse, is the principal institutional focus for that aspiration. To determine what this does and does not mean, it is necessary to do at least two things. First, one must understand the institutional status of education in a given society. What do policy makers, especially relevant officials, mean by educational policy? Does consensus exist as to the purposes of education? What are the likely trends here? An examination of these questions in the American context is the focus of Chapter 7. My conclusions are sobering. All basic policy trends in education seem to me reconcilable with what I delineate in Chapter 2 of this book as a "libertarian technicist" model of the social future.

The second thing one must do to determine whether education can yield a possible countertechnicist strategy is to ask if there is anything about education itself that is essentially countertechnicist. This requires making important distinctions between education and cognate concepts like socialization, instruction, indoctrination, and schooling, and also between technicist and nontechnicist conceptions of literacy. How a society defines the nature and purposes of literacy tells us much about its official attitudes toward the moral relationships between personal participation in society and the collective quality of language use. These matters are the subject of Chapter 8. My conclusions here too are sobering. I do try to show that there is a way in which education can be understood as an essential countertechnicist activity. But I also present arguments for the view that a democratic liberal conception of society is not easily compatible with the institutional establishment of such an educational philosophy.

The purpose of this book, however, is not to generate marching orders. Rather, in these pages I hope to take some modest steps toward clarifying the connections among technicism and our various linguistic habits in the practice of sociology, education, and moral and social philosophy. More ambitiously, perhaps, I hope that these reflections may lift the debate about the effects of technology out of the mire of apocalyptic emotions. Finally, and most ambitiously, I hope that some of the ideas in this book fortify the significance of personal dignity even amidst our most collectivist preoccupations.

Notes

1. Of such literature there has been too little—though there has been some. For instance, an important and growing sector of literature goes under the title of the philosophy of technology. Much of this material deals with social and cultural processes. Some excellent examples are collected in the volume edited by Mitcham and Mackey (1972). See also their wide-ranging bibliography in this area, published as a special issue of *Technology and Culture* (1973). Also to be noted is a volume (Winner, 1977) directly related to the concerns of the present book, which was unfortunately published too late to allow careful consideration on my part of its central arguments.

2. The threat that all grounds for authority will be reduced to that of mere technological necessity is one reason for the recent spate of volumes by social thinkers on the problem of authority. This interest is often ascribed to the emergence of a mood called "neoconservatism." I find this interpretation incomplete. The dignity of the person in a technicist environment is a problem that transcends ideological boundaries. So does the issue of collective, nontechnicist standards of values, norms, and memory.

To govern our relations with men, it is not necessary to resort to any other means than those which we use to govern our relations with things; thought, methodically employed, is sufficient in either case.

Emile Durkheim

No one knows who will live in this cage in the future, or whether at the end of this tremendous development entirely new prophets will arise, or there will be a great rebirth of old ideas and ideals, or, if neither, mechanized petrification, embellished with a sort of convulsive self-importance. For, of the last stage of this cultural development, it might well be truly said: "Specialists without spirit, sensualists without heart; this nullity imagines that it has attained a level of civilization never before achieved."

Max Weber

Once a people installs the majority as the rulers of its life, that is to say, once it introduces present-day democracy in the Western conception, it will not only damage the importance of the concept of personality, but block the effectiveness of the personality value. Through a formal construction of its life it prevents the rise and the work of individual creative persons.

Adolf Hitler

Let the public continue to regard me as the bloodthirsty beast, the cruel sadist, and the mass murderer; for the masses could never imagine the commandant of Auschwitz in any other light. They could never understand that he, too, had a heart and that he was not evil.

Rudolf Hoess
(Commandant of Auschwitz)

I would characterize [democracy, Christianity, and constitutionalism] as half truths because I do not see in them a genuine confrontation with the technological world, because behind them there is in my view a notion that technology is in its essence something over which man has control. In my opinion, that is not possible. Technology is in its essence something which man cannot master by himself.

Interviewer: And now what or who takes the place of philosophy?
Response: Cybernetics.

Martin Heidegger

Life is action and not production; and therefore the slave is a servant in the sphere of action.

Aristotle

PART I

Technicism: The Problem and Its Contexts

INTRODUCTION

The strategy of the three chapters in this section reflects my desire to avoid simply recapitulating what a large number of authors have written in the pessimist vein about technicism. Instead, Part I is my interpretation of what this literature signifies as a totality. This aim is ambitious, even presumptuous. It violates the scholarly requirement to treat every important thinker in his or her appropriate historical and intellectual context. Yet the effort seemed necessary to make. There are persons, reasonably disposed to listen yet not wishing to spend years of effort in specialist reading, who want to know whether there is any coherent pattern to these critical assaults upon modern technopolis. Part I presents one version of the critics' case.

Chapter 1 examines a number of orientations toward technicism as a moral problem and, from among them, selects the version on which this book rests. This choice reflects my sense of what is most causally significant in any final loss of control over the kingdom of things: forgetting the distinction between language appropriate to things and language appropriate to persons.

Chapter 2 attempts to accomplish three tasks. First, the chapter is designed to present an integral picture of what recent trends look like to pessimist critics from a variety of ideological perspectives. This picture is meant to be neither an intellectual history of recent times nor a history of sociology nor my personal assessment of the modern era. It is, rather, a brief theoretical sketch of what pessimist prophecy regarding the emergence of a technicist society looks like to me when this prophecy is treated as an interpretation of the relationship between social and intellectual development. The theoretical perspective that underlies this strategy is the sociology of knowledge.

The second aim of Chapter 2 is to locate the identity of sociology in this sketch. Sociology is at the same time a source of pessimist prophecy and an object of pessimism in itself because of its alleged status as midwife to technicist social controls. In Chapter 2 I try to show this dual moral significance of sociology to be the basis of sociology's civil war of conscience. This is a war often misread by outsiders either as mere squabbling between partisans of diverse jargons or as an immature preoccupation with the notion of "intellectual crisis."

Finally, I have tried to write Chapter 2 in such a way as to suggest why it could be the case that many people want to displace their freedom of agency unto "experts" of all sorts, thus facilitating a voluntary transition to a fully technicist society. When the moral and cultural categories by which people interpret themselves and their world lose all congruence with the social factual trends going on about them, a crisis of confidence in one's own resources emerges that can render people confused and dependent. In a scientific age, this dependence can take the form of ceding one's self-control to those who promise "scientifically" to alleviate the torments of incoherent linkages among intentions, actions, and consequences. Chapter 2 also introduces reference to the scientific model that most strongly tempts its proponents with visions of a technology of social survival and control: cybernetic systems analysis.

Chapter 3 moves to the center of what I see as the moral problem raised by the pessimist critics of technicism: what is it about the concept of human dignity that is not captured in an ethic formulated primarily in the interests of social system survival? No modern conceptualization of human dignity is sufficiently refined beyond the level of pious rhetoric to serve as a basis for a critique of a predominantly systems survival ethic. There are important reasons for this lack. Chapter 3 discusses these impediments in detail. It also attempts to develop a formulation of dignity adequate to surmount them and detailed enough to illuminate the reasons for regarding every normal person as a responsible agent. In this analysis, language is highlighted as the major instrumentality of personal participation in the construction of the social world. The implications of this situation are explored for a moral philosophy of personal existence in a technological society.

The Problem

From Plato's disgust with the utilitarian uses of Sophist rhetoric, through the image of Frankenstein, to the computerized monsters of contemporary science fiction, Western man has feared the products of his own technical abilities. Perhaps this is because our nightmares have regularly alternated with dreams of emulating, defying, or even replacing the gods. Be that as it may, there are many who would not consider the effort to analyze such fears worth the time. Terrors, they might say, are part of the primordial human condition and, even if exorcised, will merely reappear in other forms. This is a superficial view. Apart from the fact that human progress, in whatever form, is always associated with a confrontation with, and not the avoidance of, the demonic, there are aspects of modern times that are genuinely new and that require appropriate advances in the quality of popular awareness. The potentialities of modern technological systems are not repetitious of the past; neither are the instruments of modern warfare, the complexities of modern social organization, or the techniques of modern social surveillance and control.

Many social scientists with a professional self-image are loath to treat seriously a topic such as the demonic view of technology. One obvious reason is the vulgarity of much popular rhetoric surrounding such concerns. The vocabulary of dissent includes an arsenal of pat phrases such as "cog in the wheel," "oppressive system," "alienation," "liberation," and so forth, not to speak of more directly irrational attacks upon the integrity of the intellect itself. The temptation is great to resist such rhetoric with equally simplistic arguments for not taking it seriously. Three such simplistic counterarguments might be called the déja vu, the messianic, and the nominalist approaches. The first dismisses dissent with the non sequitur that fears have been voiced for generations and society has survived; hence, it may be expected to continue doing so. The second response reverses (and dismisses) the pessimist argument by pointing to the progressive possibilities of technology. And the third response reduces the problem to semantics by arguing that demonologies, like religions, are just dramatic names for the prosaic processes of human species adaptation and survival in nature. Fears and aspirations come and go; the human race goes on.

There are also more subtle reasons, however, for the relative neglect on the part of social scientists of the literature of antitechnological pessimism. The very idea of objective social science is brought into question by much of this literature. Trying to speak in social scientific terms of an attitude that scorns the very possibility of a vocabulary of objectivity no doubt appears to many as a self-defeating enterprise. Furthermore, sensible people are understandably hesitant to tackle the many facets of such literature, which often makes reference to cultural history, metaphysics, etymology, and aesthetics. Indeed, such a preoccupation may appear unjustifiably abstract during a time when morally sensitive thinkers are concerned with more immediate and seemingly more tangible problems—like famine, unemployment, and war—which, it is hoped, technology will prove helpful in solving. Given the faith invested in the pragmatic benefits of instruments devised by human beings for the solution of human problems, it can indeed appear as unforgivable sophistry, in an age of crises, to ask whether or not we have something to fear from these instruments themselves.

Yet the fear of technology not only persists but has found new and widely ranging expression within this century. Such expression ranges from serious philosophical and sociological literature and anti-utopian fiction to counterculture youth movements. At the heart of this pessimistic reaction is a general conviction that there is something about technological norms and values that is capable of eroding or destroying nontechnological norms and values. Whatever the final verdict, a technological civilization cannot ignore whatever substance there may be to such an assertion. Modern technology, social and physical, is humanity's most effective innovation because it seems to extend human will over nature in ways that immeasurably exceed what is possible on the basis of human anatomical capacities alone. Can it seriously be maintained that there is something about modern technology that comes to dominate persons and transform them into functions of technology? If so, it would have to follow that the vocabulary of free will, stressing as it does the mastery of instrumental means to achieve humanly defined ends, must be abandoned as illusory. Such a vocabulary would have to be replaced by a mode of discourse that equates human action with the outcomes of technological determinism and defines any serious departure from this equation as useless utopianism. The ideological rift that has emerged between partisans of different positions on these matters promises to supplant the more traditional antagonisms of "left" versus "right" (McDermott, 1969; Hampden-Turner, 1970:303–347).

There are some characteristics of the current literature of antitechnological social criticism that make precise definition of the issues at stake rather difficult. I shall cite three.

One characteristic is that much of the literature on the moral problems of technology is cast in an almost demonological style. That is, technology stands forth as a specter—something so seductive and autonomous in its influence as to abort nontechnological styles of thought and modes of experience. In some quarters this attitude has generated a mood of revolt against the whole cultural and social fabric of modernism. Over the past century or so, this revolt has ranged from ideologies of nostalgia for medieval ways of life to outright condemnations of logic and science. This apocalyptic mood is often made to stand in place of a clear statement regarding just what it is about the sociocultural organization of techniques that is capable of leading to such demonic results.

There is a second source of difficulty for those who would be precise about the moral philosophical problems raised by technology. The more subtle literature on this subject is not antitechnological in the sense of opposing physical machines as such; that is, we are not dealing merely with a critique of the social and economic effects of physical technology. Rather, there is something far more basic at stake—which, in this book, I shall refer to not as technology but as *technicism*. One of this book's intentions is to impart a reasonably clear sense of the difference between these two terms. Problems abound, however. The moment one moves beyond the direct effects of physical machinery, the distinction between technological and nontechnological norms and values becomes rather cloudy. *All* human enterprises involve techniques of some sort. There are techniques for encountering the sacred; techniques for resolving moral conflicts; techniques for creating objects of beauty. Is there any substance to the distinction between technology and technicism? Or are we dealing only with rhetoric used by some who wish to criticize whatever they do not happen to like about a technologically oriented society?

This question might easily occur to the reader who took at face value a third characteristic of these critical writings: the extreme variety of their ideological sources. Anyone who believes that the antitechnicist vocabulary reflects more than a convenient rhetoric must prepare to search for a common pattern of theoretical claims not definable in terms of currently established ideologies. It is my task in this book to locate this common theoretical pattern without doing violence to the important differences that otherwise divide thinkers of antitechnicist persuasion.

Technology versus Humanity: Some Forms of Anxiety

The general claim that technology is associated with social problems of crisis proportions needs no defense. The question of how the issues surrounding this claim should be conceptualized has generated much more controversy than the basic generalization itself. Let us distinguish four different, though obviously overlapping, ways of talking about the interaction of technology and humanity. It is possible to present each of these approaches in a manner that makes some sense of the notion that technological norms and values can erode nontechnological norms and values.

SOCIOTECHNICAL DETERMINISM

One may hold that physical technology, in the social form of the modes of production, and their appropriate social relationships, creates preconditions for patterns of social organization. These, in turn, determine the forms and contents of the more abstract world of ideas and intentions. This is the founding assumption of Marxism. In a rather loose sense this view of the importance of physical technology has become part of the conventional wisdom of general sociology. The assumption shows up, in varying degrees of alleged determinism, in the thought of man otherwise quite different in intellectual orientation, such as William F. Ogburn, Leslie White, and Marshall McLuhan. The general relationship among technology, social forms, and ideas is supposedly not a recent phenomenon, however, but a basic and persisting historical pattern. Yet much of the pessimistic literature implies that something new and more ominously deterministic has entered the picture in modern times. This pessimistic turn has affected thinkers within the Marxist intellectual orbit as much as those outside it.

TECHNOCRATIC ELITISM

One may also think of the triumph of technological norms and values in terms of the direct control of society by an elite of technicians and scientists dedicated to standards of efficiency dominated by the dictates of machine production and social bureaucratization. There can be little controversy aroused by the claim that technical experts now play a very important role in the councils of power. But to conclude from this situation that a government by science exists is quite another matter, and the major critics of technicism do not actually argue that such is the case. The following quote seems a reasonable generalization of contemporary trends as regards the interaction of politics and science.

> Even assuming that the role of the politicians will be taken on by scientists, this does not settle the question whether they will act, in the political capacity, as scientists. . . . What is likely, on the contrary, is that the scientist will be put to use by the politician, and that the technical skill of the former will prove very useful to those who know how to master not knowledge but men (Sartori, 1965:408).

This observation directs our attention to where it properly belongs: the contexts and conditions for freedom of choice with reference to those decisions that affect the collective destiny of large populations.

TECHNOLOGICAL DEPENDENCE

Accordingly, a third way in which one may speak of the effects of technological norms and values is to point to the impact, upon human freedom of action, of the exploding complexity of technological systems as such. Here the argument would be that such technological systems increasingly transcend the capacity of their directors to control them, basically because their directors do not understand them. Yet these systems perform socially desired economic functions, and hence human direction of technology is gradually supplanted by human adaptation to it.

This form of argument is very persuasive. However, upon reflection it seems to beg certain questions. Technologies cannot "make" men do anything. People act or fail to act on the basis of their interpretations of the world around them; interpretations embodied in language, institutions, artifacts, and social organization. The technological world created by human innovative effort reflects human assumptions, values, desires, and aspirations. This point is sometimes answered by adding to the argument at issue a theory of interests. Technology, it is held, is so foundational for collective human interests of all sorts that to abort it in any way contradicts all interests and would therefore be "irrational" for virtually everyone. But even with this amendment, the argument fails to account adequately for these observations. First, as the arms race amply illustrates, controlling and restraining our technologies is just as much in our collective interest as are the uses of the technologies. Second, there is general agreement about this point among almost all persons who reflect on it. Therefore, one cannot explain the absence of such commonly desired controls by means of a (at least simplistic) notion of interest conflicts. Third, a great deal more is known about how technology can be controlled than is actually put into practice. Whether one thinks of overkill ratios, automobile safety, transportation policy, population control, or pollution, numerous examples come to mind of contradictions between agreements on common interest in, and capacity for control of, technologies, on the one hand, and the actual efforts made in this cause, on the other.

Important as are these three ways of speaking about the topic, the questions I have raised plus certain features of the most sophisticated literature of pessimism all point to a fourth sense of the claim that technological norms and values can erode other norms and values. It is this fourth way, which I intend to call the problem of "metaphorical dominance," that it is the task of this book to describe and help put into a social scientific framework. Before turning to this directly, I must pause in order to explain more fully what I mean by saying that some of the characteristics of the most sophisticated literature of pessimism imply the strategy adopted in my book. It is these characteristics to which I desire to give more integral coherence in ways that do not foreclose the redemptive possibilities of science and technology in human affairs.

TECHNICISM AS AN IDEA IN THE CRITICAL LITERATURE

Several themes are notable in the best of the critical literature being directed against modern technological society.[1]

First, what is being attacked in these writings, as noted earlier, is not physical technology. Nor is it a whole society or any specific institutions, roles, or norms within it. Rather, the focus seems to be on one sector of cultural meanings: those that legitimate institutions by defining the purposes for the sake of which institutions are now organized the way they are. Technicism, in other words, appears to its critics to be a method of legitimation reflecting a particular kind of world view. In short, technicism is alleged to be an ethos, a collective mentality.

Second, technicism is not viewed by its critics as resulting from a deliberately arrived at philosophical decision. Thus, the critics do not write as though addressing partisans of a rival theology. Instead, they imply that technicism is a phenomenon comprising unconsciously taken for granted assumptions, not a set of principles consciously accepted. This does not mean, of course, that consciously technocratic intentions are not present in society. Rather, the alleged operation of a technic-*ism* as a total moral outlook is not regarded as explicit in the way other moral logics are that are associated with Western civilization (such as the forms of Christianity or constitutionalism). Technicist man is apparently not expected to be aware that he is such. This makes the diagnosis an "observer's model" on a high level of abstraction. To its critics, the technicist mentality, it seems, is a state of inferior being. It is a rejection of some essential way of being human, even a state of collective amnesia in this regard.

Third, the varieties of serious criticism I have reviewed for this book are not to be confused with any simplistic, neoprimitivist rejection of civilization. To be sure, the popular books by Theodore Roszak (1969, 1972) contain a very strong ideological core of neoprimitivism. The anthropologist Stanley Diamond has repeatedly stressed the persistent theme of anticivilization in his own brand of social criticism (1964, 1974). But despite these popular examples to the contrary, it remains true that there is no necessary connection between the critique of technicism and the primitivist tradition. It is not primitivism to feel ambivalent about man's capacity to extend his collective will into nature without risk of "dehumanization," especially if this will is clearly inspired by a motive toward the domination of nature. What the specific sense of "dehumanization" might mean, however, varies among the critics. That is, what ties the critics together is not agreement about a precise definition of what it means to be fully human. Rather, they share a common vision of technicism as an enemy of their particular notions of what being fully human is all about.

Fourth, there seems to be a shared conviction among these critics that a fateful turning point occurred somewhere in the early history of modernism. Many of them regard modern physical science, social science, philosophy, and the arts as all implicated in the problem of technicism. Technicism thus becomes an aspect of a total world view; a profound, unconscious, historical and cultural process rather than an individual or collective intellectual decision on the conscious level. This notion of technicism as a total world view should be distinguished from another, more common, form of technicism. We may distinguish these two senses by the terms *pantechnicism* and *parochial technicism*. Parochial technicism is the common state of mind in which the world is viewed myopically through the lens of some particular technique and its interests, agents, and organizations. Pantechnicism occurs when the entire world is symbolically reconstituted as one interlocking problem-solving system according to the currently dominant technological language of control. My primary concern in this book is with contemporary pantechnicism.

The conceptual center of pessimism about pantechnicism seems to be an implied general theory about the emergence of phenomena that constrain the individual's capacity to participate by intention in the direction of his destiny. Here we find a basis for the distinction whose importance was urged on the reader earlier in this discussion. That distinction was the difference between the meaning of technique in the universal sense of instrumentality and technique in the context of technicism. A simple instrumental orientation pictures human agents using techniques to manipulate some aspects of the world in the service of humanly defined intentions (love, domination, salvation, beauty, craft). Technicism is an alleged mental environment in which the vocabulary of agents and intentions has been displaced by a world picture of self-regulating objective processes. These processes supposedly emerge, out of nonteleological events of natural selection, as natural systems whose "ends" are the effective biological adaptation and expansion of the human species in physical nature. In such an environment, the language of consciousness, purposes, and free will is destined for obsolescence along with other "premodern" vocabularies like magic, religion, and witchcraft.

In strictly sociological terms the critics of technicism regard modern societies as converging in the direction of a social order dominated wholly by "techniques." This means that modern societies are supposedly becoming ensembles of social organizations whose sole reason for existence is to further the interests of humans as functionaries. Functionaries are persons who have tied their energies, faith, and hopes for personal advancement to some physical or social technology's marketability. The assumption is that the goals of techniques, for reasons that vary with the particular critic, get lost in the shuffle. All that counts is the self-perpetuation of organized technologies whose original purpose was to solve a particular problem or to provide a specified service. One example of such a broad argument is the popular view that bureaucracies typically end up existing only for themselves rather than for the fulfillment of the purposes for which they were organized. Society, in this critical imagery, is viewed as a structure of means; an assemblage of activities whose directions are not under the control of any human agent capable of setting authoritative goals in whose name physical and social techniques are definable as means. The world of techniques and its maintenance thus becomes an end in itself. Eventually, the various parochial technicist faiths become consolidated into an integral "theology" of pantechnicism. If and when this happens with finality, we will be at the threshold of a new cultural era: a postpersonal world.

In all these arguments, technicism stands forth as a malady of "false consciousness," not simply a controversial but legitimately alternative outlook on man, society, and nature. Beyond this, the interpretations of the critics diverge. These variations have their origin in the diversity of assumptions about the essence and locus of human volition, and the appropriate ends of collective action. What is shared by these Cassandras at the gates of Technopolis, whatever their other concerns, is an intuition of threat to the distinctly human status itself. The articulation of this intuition largely transcends the boundaries of earlier left-right and sacred-secular ideologies in the West.

The modern world was born, amidst the rusting fetters of manor, church, and crown, in an aura of celebration. What was being celebrated was the promise of personal participation in the affairs of society and eventually, through science, in the direction of nature itself. Nowhere was this vision expressed in more dramatic form than in the secular eschatology of Marxism. As far as actual social institutions are concerned, the greatest concrete victories of participatory democracy occurred in those societies in which the proponents of liberalism successfully propagated the ideals and practices of civil liberties. It would seem that some vast irony of unanticipated consequences is implied by the widespread presence, in these very societies, of a critical literature that now regards liberal institutions as incapable of sustaining the promise of personal and political self-determination.

Now let us return to the mainstream of our discussion. It will be recalled that we were reviewing some modes of anxiety about the specific ways in which technology can be said to be destructive of the interests of a free humanity. Three such destructive relationships were noted: "sociotechnical determinism," "technocratic elitism," and "technological dependence." A fourth conception was then hinted at: "metaphorical dominance." It is this fourth conception that appears to me implicit in the characteristics of the critical literature that was just reviewed and that this book has been designed to elaborate. What does metaphorical dominance mean? How does it differ from the other versions of the relations between technology and humanity? And what implicit theory is reflected in the phrase metaphorical dominance that can illuminate the critical themes common to a widely divergent literature of apocalyptic pessimism directed against technology? To pose these questions properly, we must exorcise from humanistic social criticsm the demonic sense of technology. As long as it is believed that technology can "possess" humanity and control its "will," a tragic and needless confrontation is inevitable between the human hand and the human heart.

Can Technology Cause Dehumanization?

Whatever it means, dehumanization is itself a human act, something done by persons. Why, then, has technology become an object of such dark visions in the twentieth-century intellectual imagination? The answer I wish to propose as the basis for this book is that three concepts have not adequately been distinguished in modern humanistic criticism. These concepts are science, technology, and technicism. My argument will not take the trivial form of substituting one for another of these as the causal culprit. Such a deterministic form of dealing with the question of dehumanization is itself part of the problem.

Essentially, technicism is a state of mind that rests on an act of conceptual misuse, reflected in myriad linguistic ways, of scientific and technological modes of reasoning. This misuse results in the illegitimate extension of scientific and technological reasoning to the point of imperial dominance over all other interpretations of human existence. That such misuse is often unintended does not make it less a human act. When threats of dehumanization lead critics to attack science and technology as causes of their misuse, such pessimism constitutes a failure of nerve as potentially demonic as the alleged sources of its inspiration. How, then, should we proceed?

As noted, in the minds of critics "technicism" is a term of derogation. I assume that no one would defend himself against such a charge by affirming it, by saying, "Yes, I am a technicist." More likely he would say, "Nonsense, I'm just being scientific." This shows the urgency of adequately distinguishing among science, technology, and technicism. Failure to do so makes it seem that science and technology are somehow determinants, in a strict causal sense, of whatever one condemns as their misuse. Not only would such criticism, if effective, demolish the benefits of science and technology by seeming to make their moral cost too high. Ironically enough, such a view would also present a picture of humanity as profoundly unfree, as incapable of mastering science and technology for human ends.

Science is controlled by certain constitutive assumptions that make it what it is. Among these regulating assumptions are determinism and a technological test of truth. Determinism assumes that if you have a complete description of the state of the world at any given moment, and a complete knowledge of the laws of nature, then any event can be predicted or retrodicted. A technological test of truth assumes that the *ultimate* standard of verification of a scientific theory is the ability to control the world's behavior by means of technological operations deduced from the theories. The scientific project, in other words, is to search for evidence relevant to the construction of deterministic models of nature. This requires a mechanistic view of the world in the sense that any explanation of a scientific sort is mechanistic. The ultimate test of understanding something scientifically is the ability to deconstruct and reconstruct the object of understanding at will.[2]

Yet I also assume, along with John Dewey but most especially with perhaps the leading contemporary philosopher of action, John Macmurray (1936, 1957, 1961), the primacy of action. That is, I take for granted that all theoretical activities are best ultimately understood as aspects of individual and collective action.[3] This all implies that technicism can operate on two levels. One level is within science itself. This occurs when technological metaphors are accepted too readily, too soon in the process of analysis and explanation. This may happen simply as a result of research laziness. More often it is a by-product of a too rigid commitment to using the same theoretical models for explaining physical, organic, and human domains of nature. The second level of technicism emerges when the constitutive assumptions of science themselves are extended to areas of discourse in which there is reason to believe they are not appropriate.

The concept of technicism, as it is used in this book, thus encompasses two acts and their consequences. The first act is ignoring, within science, the need to pay attention to the *discontinuities* between the human world and other worlds that are also the objects of scientific attention.[4] The second act is ignoring the epistemological limits of science relative to other modes of reflection and action. Like any other systematic activity, the scientific mode of inquiry is defined and regulated by assumptions that constitute that particular activity. To accept these constitutive assumptions as ultimate truths is an act of almost religious commitment. Such a commitment to constitutive assumptions is not required as a condition for profitably engaging in an activity. One may engage in an activity for moral, aesthetic, prudential or technological reasons; all these are worthy motives short of total ultimate-truth seeking. Indeed, such redemptive benefits of an activity could well be canceled by idolatrous fixation on the constitutive assumptions of that activity as ultimate forms of truth.[5]

Technicism, then, is basically a species of cognitive conquest. Many people desire a stereotypical simplification of the symbolic world. Thus, it is a historically common occurrence that certain symbolic categories get to be driven as metaphors into alien linguistic terrains. Linguistic usages are connected with domains of action. Eventually, through linguistic reinterpretation, these domains of action are led as vassals into the dominion of the new metaphors and their assumptions. Technicism is a special case of this common phenomenon of cultural imperialism. That this process can occur unconsciously does not make it any less an act of persons and therefore rectifiable by better informed intentions. Technicism, in other words, is not a necessary outcome of institutionalized science and technology. It is an outcome of ignorance regarding the nuances, and the implications for action, of language itself. Because this is so, this book is focused primarily upon technicism on the level of language. My cause is not revolution against the scientists. The enemy, rather, is our universal complicity in the degredation of linguistic discipline.

The charge of technicism is more widely applied to the uses of the social sciences than to those of the physical sciences, but equivalent arguments exist there. The controlling assumptions of science, for example, exclude explanatory reliance on the will of deities. When scientific explanations are used to "prove" the nonexistence of divinities, those who do this are sometimes accused of thinking of the world as nothing but a "thing" or a "machine." Despite appearances, it is not really science that is being attacked here but an illegitimate extension of science's regulating assumptions into all possible facets of existence.[6] In the social sciences, the issue of technicism only indirectly takes the form of debates about religious topics. Directly the issue seems to turn on certain aspects of the age-old debate over freedom versus determinism and on the related question of whether the human status is an ontologically distinctive one in the kingdom of nature. Those who have formulated the pessimist literature about technicism are, by implication, worried about the status of freedom (however they may understand the term) in a human world that they see being treated as if it were a determined object amenable to all sorts of technological interventions. Critics worry about this both because they feel this trend violates nonscientific ways of understanding the world and because they regard an all-out technological attitude toward the world as totalitarian in practice.

As regards science, my concern with technicism in this book is limited entirely to social science and social policy, most specifically, the role of sociology in each. The pessimist visions of modern culture, I wish to argue, can be integrated into a singular coherent form. This can be done by citing four directions in which the regulating assumptions of science are being illegitimately extended, to the detriment of respect for human free will.

These four extensions affect the concepts of the social *world* and of human *goals*, *problems*, and *knowledge*. Let us look briefly at each in turn to see how this book is organized to deal with them.

1. Sociology, for the purpose of scientific inquiry, views the social world as an object with its own "laws" (or "lawlike regularities"). Because of its scientific intentions, sociology conducts its inquiries with minimal reference to the creative agency of persons.[7] Humanistic criticism implies that when social science assumptions (of determinism and the prediction-control test of truth) are extended beyond the bounds of science, they can become the controlling assumptions of other domains of the social world. To the extent this is so, a technicist culture is in the making.

2. In science, "goals" are sought-after answers to research questions that take the form of specified and operationalized statements about measurable phenomena. At least in principle, scientific methods stress the virtues of technologically clear steps of inquiry. When this conception of goals is carried over into all domains of experience and discourse, then all goals are thought of as reducible to technological means of address. So it is, for example, that the National Aeronautics and Space Administration, whose activities are necessarily constrained by the technical logic of moving physical objects in space, could come to be regarded by some policymakers as the paradigm for other sorts of goal oriented organizations. Critics regard such an assumption as profoundly technicist.

3. To the extent that science addresses the notion of problems at all, it tends to regard them in principle as solvable. Problems are assumed to arise from the operation of knowable laws of physical nature. These operations are such as, for whatever reason, to affect us in ways that we find distressing or perhaps threatening to our survival. A scientific approach to problems entails manipulating laws so as to remove or alter the causes of these problems. The relations between the biochemical sciences and medical technology are paradigmatic of this orientation. Antitechnicist criticism implies that the word "problems" also applies to experiences that are not entirely conceivable, and certainly not always solvable, in this technological manner. In speaking of these extrascientific dimensions of problems, critics use phrases like "problems of living" deriving from "the human conditon," which cannot be defined precisely or technologically resolved. Such problems can only be "lived through in action." To forget this point is to have a technicist conception of the meaning of human problems.

4. In science, the notion of valid knowledge is constrained by science's regulating assumptions. Part of what I mean thereby is that scientific knowledge presumably can be used to generate expertise. In principle, this expertise can be used technologically to redesign parts of the world within the confines of known natural laws. Pushed to the limit, this would make policy implementation synonymous with applied science. Skinnerian proposals for societal redesign policies through mass applications of behavior

modification techniques are an extreme example of what such an orientation is like. Critics feel that total restriction of the notion of "knowing" to a strictly scientific epistemology renders all other conceptions of knowing illegitimate (i.e., fated to be regarded as "mystical," "imprecise," "utopian," and the like). It would follow for them that the honored desire to apply reason in the form of knowledge to one's problems must eventuate in technicism. This is true by definition so long as the concept of reason (and of knowledge) is officially limited to the sense of scientific expertise.

Given these risks, how can a critique of technicism be made to stop short of a total rejection of the very idea of science (especially of social science)? The only way to draw such a line, it seems to me, is to interpret science itself as part of the history of human freedom. In this light, there are some important justifications for a social science.

Social science is of great practical benefit in the search for patterns of apparent determinism in human activities. After all, not all activities are based on the full use of human capacities for consciousness, reason, and free agency. Science is the effort to overcome barriers to the more refined uses of these capacities.

Social science has value for human self-discipline in an evolutionary sense. Its ideal norms teach humility, respect for the complexities of evidence, and cooperative endeavor. Its findings provide human aspirations an anchor in the obdurate realm of factual existence. Such an anchor tempers wishful thinking with the need for discipline, patience, and planning.

Social science is part of the larger quest of art, religion, and philosophy: the search for apparent order in the world on whose foundations the terrors of chaos may be held at bay.

Social science yields important technological benefits that can be used as tools for the more efficient fulfillment of consensual human ends in nature and history.

If we fail to distinguish, in our thought and action, among science, technology, and technicism, one of two things will happen. We will eventually demonize science and technology to the point of some great religious convulsion of primitivist simplification. Or we will use science and technology to make of ourselves God, creating our descendants in our image. Of this possibility C. S. Lewis once said,

> In reality, of course, if any one age really attains, by eugenics and scientific education, the power to make its descendents what it pleases, all men who live after it are the patients of that power. They are weaker, not stronger: for though we may have put wonderful machines in their hands we have pre-ordained how they are to use them. . . . There is therefore no question of a power vested in the race as a whole steadily growing as long as the race survives. The last men, far from being the heirs of power, will be of all men most subject to the dead hand of the great planners and conditioners, and will themselves exercise least power upon the future (1965:70)

Notes

1. The works examined as the basis for this book's analysis of pessimist prophecy were selected from among the most intellectually significant of that genre of writing. There is a voluminous list of more popular works of varying quality. The criterion for significance was what could be called twentieth-century versions of theoretical fundamentalism. That is, I sought the best examples of criticism that revealed the theoretical and philosophical bases for the critical stance. Selected for close attention were Heidegger (1950–1951, 1967, 1968, 1969) and the major secondary work on Heidegger by Richardson (1963); Mumford (1962, 1966, 1970); Ellul (1960, 1964, 1965, 1967a, 1967b); Arendt (1958, 1963); Marcuse (1955, 1964, 1968); Hayek (1948, 1955, 1967a, 1967b); von Mises (1956, 1957, 1960, 1962a, 1962b); and Sypher (1962, 1968). These works were those that I accepted as the foundational literature of criticism from which I drew the spectrum of ideas that I have sought to integrate into a singular theoretical framework.

2. For an excellent, brief account of these constitutive assumptions of science and their historical roots see Jonas (1959).

3. My whole purpose in this book could, in a nutshell, be said to be the extension of Macmurray's philosophy of persons and action (1936, 1957, 1961) to the critique of technicism in a variety of linguistic domains.

4. In this book, concern about technicism within science is reflected in my criticism of prematurely mechanistic reasoning about sociological concepts. Concern about technicism as the illegitimate extension of science is reflected in criticisms of discourse in the areas of moral philosophy, social policy, and education.

5. This comment makes me sound more relativistic than I personally feel about scientific evidence and truth claims. I regard the scientific project more disciplined with respect to canons of evidence and the constructive effects of systematic doubt than any other mode of inquiry in human history. As such, in my judgment it has higher claims on our sense of credibility than any other product of the evolution of human consciousness.

6. This point has been heavily obscured by the recent mindless dispute over "creationism" versus "evolution" in the American public school curriculum. "Creationism" is not a theory in any recognizable scientific sense. Nor is it the sine qua non of a religious (even a Judaeo-Christian religious) view of the world. After all, "God works in many ways his wonders to perform." For devout persons, one of these ways could well be biological evolution. The point of confrontation between science and religion is not any one theory. It is, rather, the extension of scientific metatheoretical assumptions to a point of dominion over all other metatheoretical assumptions about the world.

7. In the free will tradition of discourse, "creative" refers to the capacity of persons for unpredictable action, for the constituting of new meanings, and for the formulation of new intentions that causally affect the social world. In short, "creative" means the capacity of persons to help make history. In the context of the notorious "Cartesian problem," either persons are the unfree, determined "objects" of scientific study or they are the free but passive spectators of their own behavior. For an extensive development of the vocabulary appropriate to a view of persons as "agents" see Macmurray (1957).

2 Dark Prophecy

THE IRONIC IMAGINATION AS SOCIAL THEORY

Pessimist prophecy is an exercise in the diagnosis of crisis. The form that it takes is often the detection of ironically malign consequences of virtuous intentions. In this chapter I try to do justice to such imagination by means of an effort to organize a wide variety of specific anxieties, nostalgias, diagnoses, and criticisms about technology into a coherent framework. However difficult the task, it is vital to attempt it. The point of the effort is to place technicism into a developmental, cultural-historical framework lest it appear simply as a form of spiritual apocalypse or a fateful denouement of human nature itself. If contemporary capitulations to technicism can be rationally explained as unforeseen but reversible outcomes of more noble intentions, then the sense of demonic inexorability that so often accompanies the ironies of history may weigh a little less heavy on the heart. Such, at any rate, is my hope for this chapter.

I begin with a brief commentary on the concept of crisis. This term is currently in wide circulation. An introduction to how the concept of crisis is generally conceived in this book may be helpful in understanding what follows, with more specific reference to why some regard technicism as a form of crisis. After that I shall turn to the main task in this chapter: the elaboration of ways to explain the emergence of technicism as a developmental process.

Social Crisis and the Limits of Reform

The term "crisis" is used too loosely in popular discourse. I think it should be used only with respect to problems one regards as no longer amenable to social reform. What does it mean to say this?

18

Social institutions are legitimated both by popular ideologies (popularized notions of what institutions are good for) and by theories (formal, supposedly scientific accounts of institutions' origins, functions, and consequences). Why are ideologies and theories considered important enough to be worth applying in practice? Because they are thought to reflect shared values of a fundamental but abstract sort about what sort of society is worth living in and what goals are worth living for. Thus, in the modern liberal world, broadly accepted values like "freedom," "personality," and "opportunity" legitimize notions like consumer sovereignty, competition, hegemony of market criteria, separation of parochial loyalties from the larger organizations of society, separation of political power and economic pursuits, social pluralism, and statutory civil liberties. Under the assumption that institutional ideologies reflect the values constitutive of the culture itself, social elites are expected to implement these ideologies through institutional practice. If things go wrong, the public then assumes that theory is not being translated into practice. Efforts are undertaken to "reform" institutions in line with their legitimating ideologies. Antitrust laws and various current efforts to insure participatory democracy may be cited as examples from contemporary American society.

The concept of reform, then, is based on the assumption that, with a little tinkering or some new mechanisms (laws, procedures, techniques), existing institutions can be made to satisfy popular expectations. What if this is not so? Then a true crisis exists. A crisis should be defined as a situation in which institutions prove irrelevant to popular expectations—irrelevant in the sense that reforms to make them work according to their legitimating ideologies are beside the point. How can this happen?

First, there might be a drastic shift in the basic values shared by the population as a whole. For example, a shift from habits of frugality to mass consumption based on easy credit may represent such a value shift for Americans. More decidedly, however, a large-scale rejection of democracy itself or of technology, or a collective demand for authoritarian religion, or a mass rejection of job employment as the main criterion of social independence would be examples of a truly basic shift in American values.[1] If this shift in popular expectations and values were not recognized by social elites or were to prove unassimilable by existing institutions, then massive strains would build up in the fabric of the social order. These strains would not be resolvable by the reformation of existing institutions in the direction of their ideological models because the assumed relationship between these ideologies and the values they were supposed to reflect would no longer be valid.

It may also happen, however, that basic values do not change but rather new conditions arise that make existing institutions irrelevant to the persisting values they are supposed to serve. Such new conditions can be quite varied. For example, massive changes in physical technology can create entirely new situational circumstances. Or unique problems may derive from unanticipated consequences of existing institutions, consequences such as changes in the population growth rate or in the pace of social change, new organization complexities, and disjointed trends of economic development. Other sources of new conditions are political events such as war, revolutions, and novel patterns of political participation. Finally, psychic changes outside the realm of values are possible; among them new experiences of relative deprivation brought on by mass communications or demands for new goods implied by existing values that had not seemed satisfiable at an earlier stage of technological development.

In the case of either changed values or altered conditions a social crisis exists. Subsequently I shall argue that in liberal societies the real world conditions intervening between institutions and popular expectations have so changed that the institutions in question can no longer work in the way that their legitimating ideologies lead people to expect. Legitimate liberal institutions are in a state of crisis that cannot be eased by intellectual fiat or simple moral exhortation.

The outlines of this crisis were understood by a number of social thinkers in the nineteenth century, conservative as well as radical. To such thinkers it was becoming evident that industrial civilization was going to be characterized *simultaneously* by unregulated concentrations of vast power *and* centrifugal tendencies arising from the social division of labor. This awareness precipitated a renewed search for the foundations of social unity and that search constituted much of the origins of modern sociology. It is not very instructive to view sociology simply as a chapter in the history of science if one means by this characterization only the increasing utilization of scientific methods in the study of social behavior. This rather textbookish approach seems to imply that the metaphysical and messianic passions of nineteenth-century social thinkers were nothing but "unfortunate departures" from their otherwise "scientific" contributions. Sociology is better viewed as a language whose primary mission is to articulate the quest for a specifically modern understanding of the ancient notion that man is a social being. This quest had been made necessary by both the great social changes attending the rise of modern science and industry and the effects upon the intellectual imagination of themes like atomism, egalitarianism, and materialism rooted in the great philosophical revolution of the seventeenth century. The sociological thinkers of the nineteenth century do not divide into categories like scientific, messianic, and metaphysical. Rather, they divide over various debates. One such controversy centered on the relative significance of deliberate human action versus spontaneous historical processes for the resolution of industrial civilization's emerging contradictions. Another debated question was whether one should look to a romanticized past or a utopianized future for the baseline of criticism against the present. Finally, many of these thinkers were also preoccupied with the redemptive possibilities of converting popular consciousness to new modes of historically progressive thought. Thus, Comtean positivism and Marxist humanism were intended to be states of redeemed consciousness as well as proposed modes of social reorganization.

Before pursuing these points in relation to technicism, we must first consider briefly the emergence of secular modernism and its characteristics. It was within this context that the classical liberal understanding of personal agency and self-determination took shape.

Modernism: The Foundations of Irony

"An architectonic vision is one wherein the political imagination attempts to mold the totality of political phenomena to accord with some vision of the Good that lies outside the political order" (Wolin, 1960:19). Of all the upheavals of history and culture, it is difficult to imagine any of greater significance than the decline and fall not of some one vision of the good but of the good itself. The rise of the notion that there is no such phenomenon as the good in the objective nature of things must be the most ironic possible anticlimax to centuries of bitter conflict between those who felt themselves empowered to define the good.

Many volumes have been written about the premodern worlds in which men could still look "outside" into nature for clues about the truths and traps of the passions "within." Nonetheless it remains difficult for us today to appreciate how much in those premodern times could hang on a turn of priestly phrase, a philosopher's syllogism, or a footnote to the scriptures. There is a way in which the sacred words of modern times ("freedom," "equality," "toleration," "individualism") and the institutions based upon them ironically echo the void, a suspicion that their existence bespeaks not some triumph of human will or insight but a simple measure of the total indifference of nature to all moral conflicts.

In general, this point is on the way to becoming part of the conventional wisdom although its scope and implications have not been widely appreciated. For example, it is not a matter of general awareness how changes in the theories of physical nature were linked with the destiny of moral, political and social doctrines. The average modern man is given little chance to appreciate the details of why the great transformation in the theory of motion dating from Galileo had also to constitute a moral revolution. Nor will such a man spontaneously understand that the monetary interest he pays without question for a loan at the bank was for centuries a topic of bitter conflict. He is likewise unaware that the physician he treats with such deference was once not as highly regarded because the medieval hierarchy of knowledge placed medicine somewhat low due to its lack of direct concern with contemplation of the divine. The physician was seen as concerned with more transitory and material, hence subordinate, truths.[2]

None of this is to imply that most people, modern and probably medieval alike, would not prefer modern times. Statistically, life in the modern West is generally safer and certainly more personally promising. But that is not the point. Our concern is with the consequences of those massive shifts in meanings that are today defined as the transition from medieval to modern culture.

The architectonic theme in medieval life was perhaps nowhere more effectively symbolized than in what Benjamin Nelson has called the continuities among "conscience, casuistry, and the cure of souls" (1965a). The dissolution of these links is an appropriate indicator of the demise of a world view.

Although the individuality of men was certainly recognized, the medieval experience of conscience was a datum with an irreducibly objective referent, namely, the aspect of divinity in each man that linked him to God. The moral logics that reconciled the varied dictates of personal conscience into the great structure of authoritative revelation and legal traditions were called casuistries. And for the spiritual conundrums and suffering deriving from the problems of conscience and action, there were the technics of the cure of souls (Kirk, 1927; McNeill, 1951). What the metaphysical significance of the modernist revolution partly entailed was the disjunction among these three spheres: conscience became gradually reinterpreted into "interests," a purely subjective phenomenon, the pressures of personal desires and utilities; casuistries were transmuted into "ideologies," techniques for the justification of power and interests (J. A. W. Gunn, 1969; Holtman, 1950; Mannheim, 1936); and the cure of souls became, in time, a technical problem of secular medicine.

A second major change heralding the demise of the medieval world order was the decline of epistemological emphasis on Christian revelation and its associated ecclesiastical controls as reference points for secular authority. Conflict between church and state was a medieval as well as a modern phenomenon. But from the Reformation onward, theological individualism, followed by political, economic, and social individualism, became a part of the modernist attack upon all received traditions of corporate authority. Modernist materialism in scientific philosophy gradually eroded the metaphysical foundations of divine revelation as a source of guidance for social control.

The third great change signaling the end of the medieval perspective was the emergence of a new view of nature. To the men of the medieval period, nature appeared to be divinely ordered. To the modernist mentality nature appeared indifferent in the sense that natural laws seemed irrelevant to human moral aspirations. His passions forced back into the relativistic domains of his subjective psyche, modern man became more and more of a spectator of an objective world to which he could respond only with contemplative passivity or with technological activism.[3] Modernism saw the rise to commonsense status of the great Cartesian dualities of "mind versus matter," "material versus spiritual," "subjective versus objective," and "individual versus society." In part, as we shall see, this was because these dualities provided a rationale for whole social institutions. In time, these trends toward atomistic individuation and the relativization of conscience and morality congealed into conditions perceived by nineteenth-century social thinkers as threatening to the sense of a common *humanitas* in the West.

The broad features of the modern scientific world view since the seventeenth century have been increasingly materialistic, value-nihilistic, and technological. That is, modernism is materialistic in metaphysical orientation (Lange, 1925). Modernism is value-agnostic to the point of nihilism as regards the possibility of grounding collective standards of values in anything more transcendent than the simplest shared utilities (like power, wealth, and security) of one's immediate personal circumstances.[4] Finally, modernism is technological as regards the appropriate stance of the human species in nature. Nature exists, that is, to be harnessed to human will.[5]

In this emerging cognitive setting, the social institutions of the liberal order gradually took shape. With historical hindsight we can now recognize some basic themes that lent coherence and legitimacy to the institutions of liberal society. These themes can now be seen more clearly as a complex of responses to the problems of social disunity and conflict that attended the demise of medieval civilization. On the cultural level these themes also proved functionally responsive to the challenges of relativism emerging as implications of the modernist world view. One of these themes was the general desanctification (*secularization*) of the world in general and of political economy in particular. Perhaps nowhere was this trend more clearly exemplified than in the total transformation of human skills and of the earth itself into commodity resources for commercial production. A second theme was the *market* principle of socioeconomic organization. A third theme was the principle of toleration that eventually evolved into the positive notion of *pluralist representation*. These terms bear some elaboration.

As the sacral interpretation of nature and of medieval social institutions declined, social freedom began slowly to take on its contemporary sense of unrestrained egoistic atomism and individual pursuit of economic gain. With this change, the notion of the market commenced its rise to central importance in modern life, culminating in the ideology of laissez-faire. The market in its general sense came to be conceived as a structure of arrangements for the optimally efficient negotiation, through market dominated prices, of transactions between legally defined persons engaged in the free pursuit of economic advancement in civil society. Thus, freedom came to mean liberation from all traditionally ascribed constraints upon such pursuits short of self-evident violation of the similar freedom of others (such as chattel slavery).

In this context the meaning of politics began to change. Instead of being a means of implementing (or at least protecting) common values and traditions, politics, in the shadow of the market metaphor, eventually became a methodology for the maintenance of public order and the enforcement of voluntary private contracts. In nonliberal societies, of course, politics continued to be defined as responsible for more than this. Except for those espousing outright fascistic or theocratic ideologies, however, even activist political regimes came to justify their operations in the name of the individual and his interests. In the eyes of their apologists the communist societies are no exception since they postpone the liberation of the individual into an allegedly still freer future toward which it is the self-proclaimed task of the regime to prepare the way.

Basically, liberalism is an institutionalized recognition of value agnosticism. The great religious and class wars that followed the close of the middle ages were, among other things, moral crusades by partisans of rival notions of the true and the good. The modern world view can present no official solutions to such conflicts because its dualistic categories lead inexorably to relativism. Yet, liberal democracy has managed to carry forward the program of equalizing social, economic, and political opportunity to a point unparalleled in history. It has done so without pausing to resolve ontological problems of value relativism and skepticism. This was possible, first of all, because of a profound shift in the public definition of values. Despairing of the possibility of ever proving what was held sacred to be rooted in the objective order of nature, the modern mind increasingly fell back on the criterion intelligible to all, even when its content was variable. That criterion was utility. The modern analogue to the medieval hope for a human community of the soul became the society of utilitarian rationalists. Whether expressed in egoistic or social utilitarian terms, most modern theorists of political economy laid their doctrine on the altar of utility even as their predecessors offered theirs up to the scrutiny of divine will.[6]

Unlike the corporate hierarchical ideal of medieval social life, liberal societies regard socioeconomic strata, ethnic groups, religious parties, and other value partisans as arranged along a morally neutral spectrum of equal interest groups. These groups are allowed to struggle for the right of representation in the councils of power, a right justified now by the general doctrine of pluralism. In the imagery of pluralism the great historical conflict groups of conscience and ambition are transmuted into the status of equivalent interests. Their rights, although obviously dependent upon the might of their partisans, are at least in the confines of liberal ideology considered to have equally accessible stalls in the marketplace of ideas.

We have now to examine the question of whether this accomplishment of social progress amid ontological doubt obscured a hidden agenda in the heart of cultural modernism, an agenda whose forms are becoming recognizable in the pessimist vision of a pantechnicist world.

Technicism: The Forms of Irony

In every tradition of both theoretical and commonsense discourse, certain terminological habits evolve. These habits become symbols for affirming those features of the environment that are sensed as distinctive coordinates of social reality. Let us now examine three sets of such terminologies in modern thought. These terminologies are recognizable to persons conversant with commonsense discourse and with the type of speculation that, since the nineteenth century, has gone under the name of classical sociology and political economy. These terminologies are also axes of the major theories of social development that proved so important in Western notions of what economic development would portend for the societies and peoples of the non-Western world.

I have in mind the phrases *"traditionalism versus modernism," "economy versus society," and "communal versus associational organization."*[7] The problems associated with the processes reflected in these terms have now been globalized, thanks to imperialism and the operations of the world market.

The main theme of the argument to be developed in this section is that some of the unanticipated consequences of the transformations symbolized in the phrases just cited provide the definitions of the fully technicist society. These consequences were unanticipated precisely because, as was said earlier, what was expected was the emergence of a social order guaranteeing the fruits of personal self-determination.

There will be no review of the material and social history of the eighteenth to twentieth centuries. The well-known events of that period—massive population increase, emergence of large-scale social and physical technologies, development of devastating weapons systems, increasing rate of innovations, and unevenness of economic development—obviously form the practical backdrop of the discussion to follow about theory and social trends.

My presentation proceeds as follows (the intellectual strategy of the whole chapter is diagrammatically presented in Table 1). Each of the three developmental terminologies cited will be discussed in turn. First comes a brief review of the polarities under analysis. Then it will be shown that each polarity has associated with it a key moral problem. This aspect of my strategy can easily be misunderstood. It is *not* asserted that only three moral problems exist. Neither am I suggesting that there is some ultimate way in which the decision to label something a moral problem can be grounded in a purely scientific conception of moral truth in the manner once assumed by those who liked to employ the term "social pathology." What I do argue is that, in hindsight, it becomes evident that some (obviously not all) unanticipated consequences of any actions are experienced as inconvenient, or as

THEORETICAL STRUCTURE OF THE PESSIMIST PROPHECY

	CULTURE	SOCIETY	POLITICS
Predictions	Metaphysics of nonvolitional world of "facts"; functionalization of all thought; final fusion of information and propaganda	Servomechanistically planned cybernetic state	Elitism of technocratic and welfarist engineering
Social Trends	Decline of political vocabulary of volition and emergence of "technical" analysis of social problems on mass statistical basis.	Increasing institutional integration of society	Obsolescence of local spontaneous responses to problems of social justice and ecology.
Moral Problems	"Alienation" (Disjunction between metaphysics of volition and metaphysics of objective natural processes)	"Legitimacy" (Conflict between criteria for making private, as against necessary societal decisions)	"Solidarity" (Tension between centrifugal effects of social variation and need for common criterion of obligation to primary social community)
Intervention of Material Events	Massive population increase, large-scale technologies, weapons systems, increasing rapidity of innovation, unevenness of economic development		
Sociocultural Polarities	"Traditionalism versus modernism"	"Economy versus society"	"Communal versus associational organization"
Liberal Solutions	Secularization as desacralization of nature and of political economy	The "market" principle	Pluralist representation

↑ Decline of the medieval symbolic synthesis ↑

painful, or as downright malign. Actions can have effects that are the moral opposite of what was intended. It is in light of the *moral intentions* exemplified by any given ethos that certain social events and trends become perceived as demonic. Nineteenth- and twentieth-century social thinkers have produced a literature concerned with three major moral themes: alienation, legitimacy, and solidarity. These themes can be understood only in light of moral expectations embodied in the liberal ethos.

In the final stage of this strategy I specify the connections between the social and cultural trends currently evident in modern industrial societies, and I present the predictions that summarize the fears projected by pessimist critics about the technicist implications of these trends. These predictions form the dimensions of a technicist society.

Let us turn now to the first of the three polarities of social development.

TRADITIONALISM AND MODERNISM

Increasingly, in daily life around the world, the terms traditionalism and modernism are used less to explain anything than to express a collective sense of "no return." Of course, the context varies in mood from the elation of progress to the pathos of nostalgia, depending upon the circumstances surrounding the introduction of this polarity between tradition and modernity into the speaker's life. (The word "secularization" often serves as a synonym for modernization.) In many ways this sense of fateful change, of being refugees from the past, is generated more by the monistic materialism, the nihilism, and technologism of the modern world view than it is by the effects of specific structural forms of modern societies. As Karl Polanyi (1957) has so poignantly argued, and as has been repeatedly documented in anthropological monographs, the modern world view is introduced most sharply via the immediate, subjectively experienced relationships between a man, his kin, his land, and his work.

It is difficult to exaggerate the terrorizing impact upon all premodern notions of human status, of many common facets of modern culture.[8] The abstractions of modernist progress become very real sociologically when the graves of ancestors, the sacred groves of the gods, the streams and fields of heroic legends, and the secret places of ancient ritual alike are swept aside and replaced by a single title deed to a marketable factor of production. This point is not contradicted by the fact that modernism, both culturally and sociologically, is also associated with a new sense of personal freedom and initiative. Time is an important factor here. It is after freedom and individualism have become taken for granted, and after the modern world has been personally experimented with, that the sense of something lost often begins to set in. That this feeling usually takes the form of romanticist reinterpretations of the past should not deflect attention from the cultural importance of such nostalgia. Even the United States is currently witnessing within its borders a movement toward the communal restoration of ethnic nostalgia.

My concern is not with this story as a whole. I am interested rather in what there is in particular about the experiences reflected in the distinction between traditionalism and modernism that is relevant to the theoretical structure of the pessimist vision of technicism.

Certain aspects of the concept of secularization can be elucidated by some attention to the cultural history of the terms "theory" and "practice." These terms go back very far in Western thought (Lobkowicz, 1967). But as concerns modern and premodern interpretations of theory, the great divide must be understood in terms of modern philosophy's conclusion (one not fully articulated until almost the present century) that theory can give no foundation for public consensus on the meaning of truth in the realm of moral phenomena. Theory in the very broadest sense once meant references to doctrines and convictions about the true, the good, and the beautiful. Issues of theory have great practical significance, even if indirect, in any society that regards political intentions as related to, representative of, or based upon verifiable standards of what is true and good.[9]

In the medieval Catholic period, theory was understood to be the product of both natural reason and divine revelation. One aspect of the modernist cultural revolution has been the rejection of revelation as a valid source of knowledge. Relativistic perspectives have undermined all claims to absolute knowledge, including the once self-evident truths of natural reason. Beginning perhaps with Machiavelli, politics was gradually redefined as a secular problem in the methodology of social order. In the contemporary concept of civil society, politics is not an effort to create the good society but a method for maintaining the stability of a society so that people may pursue in relative peace the consequences of their own private understanding of what is good for them.[10]

With such changes the world of day-to-day practice comes into its philosophical own. The "practical" is now equated with the operations of the material world. The categories of moral theory are thus forced more and more into the immaterial and hence abstract clouds of vague moral "principles" (or "values").

In this kind of setting groups from different moral communities meet in the no-man's-land of interest bargaining, and the common medium of exchange becomes not moral suasion but the practical coinage of utility. The pun is deliberate. Bentham apparently understood well what the vast psychic importance of money in a utilitarian society would be.

> If, then, between two pleasures the one produced by the possession of money, the other not, a man had as lief enjoy the one as the other, such pleasures to be reputed equal. But the pleasure produced by the possession of money, is as the quantity of money that produces it: money is therefore the measure of this pleasure. But the other pleasure is equal to this; the other pleasure is therefore as

the money that produces this; therefore money is also the measure of that other pleasure. It is the same between pain and pain; as also between pain and pleasure . . . if we would understand one another, we must make use of some common measure. The only common measure the nature of things affords is money. . . . Those who are not satisfied with the accuracy of this instrument must find out some other that shall be more accurate, or bid adieu to politics and morals (quoted in Commons, 1961; p. 233).

Such a society remains stable if at least the more influential participants are convinced that the basic game is structured fairly and if the circumstances that impinge upon the game remain reasonably constant and predictable.

Both requisites have proven short-lived in the industrial phase of modern history. The fairness of the rules came to be challenged first because of an increasing sense of inequality on the part of those players who felt themselves trapped in severe conflicts not of their own making—class wars and international military encounters. Second. the objective circumstances impinging on the game, such as changing population growth rates, new resource bases, new information access patterns, new forms of labor discipline, and environmental traumas growing out of technological progress, have generated unanticipated crises not amenable to solution by the pragmatic but leisurely political processes of civil society.

The result among many intellectuals and some "men of affairs" has been a tendency to view civil problems of practical politics as disguised "technical" social problems that, once so defined, can be solved by experts. This approach, as scholars have shown, has taken many forms (Haas,1964; Sewell, 1966). However, the critics of technicism seem to fear that in the context of the relativism of modernist culture the new elites of expertise will be able to agree only on material and mathematical (positivistic) criteria of judgment. Despite profound limitations from a humanistically refined point of view, these criteria prove so impressive and baffling in their complexity to the average citizen that elites, it is feared, will be tempted to substitute for persuasion the manipulation of public consciousness. Because of ignorance about the concrete referents of politico-ideological appeals being directed at them in such a complex world, the public will be decreasingly able to discriminate between information and propaganda. The facts of any given event or state of affairs will necessarily be so complicated and interconnected with other events as to be beyond the comprehension of any single individual. Therefore, the process of manipulation will include the creation of pseudoevents, deliberately for the sake of propagandistic communication and inadvertently through the natural results of commercial mass communication. The public can be expected to cooperate because of its desperate need for a stereotypical comprehension of reality (Boorstin, 1964). Critics of course are aware that such phenomena as popular ignorance and elite manipulation of consciousness are not historically unique to modern times.

What is new as they see it, however, is the persistence of these trends in an environment whose dominant ideology includes the promise of public enlightenment through universal education, mass media of information transmission, democratization, and welfare programs designed to transform *all* dependent "masses" into independent and self-sufficient citizens. This theme has been especially stressed by Jacques Ellul (1965).

According to the pessimist prophecy, in growing contrast with the liberal notion of private interests, the whole procedure of policy legitimation will be marked by ever increasing references to the abstract but politically potent concept of the public interest. In time, the order of civil politics, with its justification by the private pursuit of happiness, will be replaced by a public order of technicist social controls under abstract rationales of social justice, political stability, and even human survival (Ellul, 1965).

Modern secularism has been accompanied by a complex of critical streams of thought centering around a moral problem that has come to be generally known as *alienation*. Two particular facets of this notion, which can be conveniently labeled "reification" and "homelessness," are relevant here.

The question of reification was raised to explicit status in social theory by the young Karl Marx.[11] Pointing to all those aspects of modern life that lead men to experience the products of their own collective work, thought, and effort as parts of an "objective" world that then rebounds as a constraining structure of "necessity" upon their own freedom and that of future generations, critics of Marxist humanist vintage raise the following question. If men have created the social world historically through their own decisions and labors, are the results of these actions always fated to be experienced by them and their descendants as chains upon their own volition? And if the works of men are to be experienced as more than this, as is explicitly promised in the Marxist notion of liberated men repossessing their history, what alterations are necessary in the way in which men participate in the world via the day-to-day structures of concrete social, economic, political, and cultural existence? Seen in this way, Marx's definition of the problem of alienation remains unsurpassed as the modern secular formulation of the ancient clash between free will and determinism.

Although the notion here labeled "homelessness" is commonly associated with the arts,[12] its import is no less philosophical and political than the Marxist critique of reification. Feelings and convictions of homelessness seem to arise when the world is scientifically defined as a materialist arena of forces and objects driven by laws irrelevant to human experiences of beauty and ugliness, love and hate, passion and fulfillment, salvation and damnation. Of course, it is not being claimed by those who argue in this manner that such phenomena are not part of the existing realities of human life. Rather, it is that the modern (scientific) world view makes it illegitimate to speak of them as "objectively" part of the world, forcing us instead officially to define such evaluative and emotional considerations as "merely subjective" projections of men's inner lives.

> Thus, religion and art lost their unquestioned birthrights in the homeland of human reality, and turned into strange messengers from the higher unreality, admitted now and then as edifying or entertaining songsters at the positivist banquet (Heller, 1959:213).

The modern world view, in other words, forces upon our consciousness a new form of the disjunction between "private" and "public" such that many things that people consider sacred for their sense of being human come gradually to be defined as random, private, and subjective. Only the calculable, the utilitarian, and the predictable elements of life are allowed the status of "objectivity." The world, once an "enchanted garden" (to use Max Weber's memorable phrase), now becomes disenchanted, deprived of purpose and direction, bereft, in these senses, of "life" itself. All that is allegedly basic to the specifically human status is forced back into the precincts of the "subjective," which, in turn, is pushed by the modern scientific view ever more into the province of epiphenomena, dreams, and illusions.

In the logic of the pessimist view, examples such as these of countervailing humanistic criticism are transitional and will lose their force once the processes of secularization have completed their course. This completion will be highlighted by three cultural events. One is the final moral neutralization of "conscience" by theories of conditioned socialization. The second is incorporation of frontier biochemical and genetic research into the polemical structure of a fully materialist account of man in nature, an account that will be used to justify direct control of human behavior by means of biochemical intervention and behavior therapies. The third development will be the social utilitarianization of all norms and values. This is a process that sociologists have studied empirically more closely with reference to religious phenomena than anything else. What is involved is the general epistemological assumption that nothing exists, or is to be related to, for its own sake. Rather, adequate explanations of any norm or value are expected to consist of an account that shows how that norm or value is "functional" (useful on a latent or manifest level) for some other goal extrinsic to itself. This orientation is simply an extension of the moral thrust of utilitarianism. A pessimistic interpretation of this process turns on the argument that it makes people increasingly insensitive to the contours of experience because the mind and heart are forever sliding around the edges looking for the "useful." In place of the desire for truth and understanding, we find the search for an indirect quality of any experience (its usefulness) raised to the level of supreme moral significance.

According to the pessimist prognosis, the *fully* secular individual is fated to be conditioned to a public world dominated only by quantifiable and instrumentally manipulable facts. Such a person will presumably experience participation in this public world only as a means in the name of ends defined behind his back, as it were, by those responsible for applying to his life the functional logic of efficient production and societal stability.

In an autocratic technicist society the individual will learn to regard other notions and styles of participation as exotic, deviant, neurotic, or utopian. In a libertarian technicist society (which I discuss later as the probable heir of liberalism) the individual will experience other other forms of participation as leisure play in a semilibertarian *private* realm of social existence. In either case, the prophecy of men collectively and consciously making their public world as a whole, a dream cherished in separate and conflicting ways by humanists of both liberal and Marxist persuasion, will be a thing of the past.

ECONOMY AND SOCIETY

Laissez-faire liberalism represented the triumph of philosophical nominalism in the realm of political economy. "Society" becomes simply a name for the products of contractual agreements between individual calculators of utility. The cultural significance of the laissez-faire version of the market concept therefore does not lie in its economic functions alone. Its importance derives also from its ideological affinity to an atomistic, mechanistic, and utilitarian world view. The triumph of the market principle was related to its apparent capacity to provide a conceptual account of nonautocratic decisionmaking processes in a Hobbesian context of potentially hostile unit egos. In the ideology of the market principle, the multigenerational "society" of shared authoritative and continuous traditions is replaced by an "economy" based on calculations of contemporary self-interest. Liberal freedom, at its root, stands for the principle of a fresh start for every individual. It promises liberation from the fate of the fathers and in this sense annihilates social time. In its place is substituted the breathless freedom of an eternal now.

It is necessary to remember that we are talking about the logic of a legitimating world view and not about the day-to-day processes of historical reality. The idea of the free economy is a historical tendency and a basis for policy rationales. Or, if one is cynical about it, market freedom may be seen as an ideological smokescreen for those who find it politically convenient to preach one thing in order to conceal its opposite. As William Grampp has written about the so-called Manchester liberals, "To look to the Manchester School for a statement of laissez-faire is to put before it a question which was not part of its purpose" (1960:2). The real-life history of the free market was a patchwork of inconsistencies such as simultaneous pressures toward free trade and domestic protectionism. The true operations of self-interest escape the consistent logic even of an ideology of self-interest.

But having said this, it must be added that although reality does not correspond with the ideological constructions of a Herbert Spencer, the logic of the market principle nonetheless remains the main organizing principle of contemporary liberal ideals of individual autonomy and freedom of choice. To abandon the logic of this principle is to move outside the ambience of the liberal understanding of social freedom. Even some thinkers highly critical of capitalism and sympathetic to the ethical vision of socialism are reluctant to abandon the operational victories of liberal freedom (Heilbroner, 1969). It is not wise to look upon such matters as theoretically abstract issues of no importance. To do so is to fly in the face of all historical evidence that ideas, however indirectly, have important consequences for the legitimation and delegitimation of social institutions. Downgrading the indirect influence of ideas, in this case the market principle, invites premature confrontation with the unanticipated social consequences of the new ideas one necessarily must put in the place of those rejected. The history of the "higher" freedoms of the "socialist" countries gives ample evidence of what this can mean in practice.

Given the inevitable struggle surrounding the market principle between its ideals and its history, it comes as no surprise that the general moral problem of legitimacy (or what used to be called sovereignty), like that of alienation, has continued in the modern world to be a major preoccupation of classical social theorists of all ideological persuasions. The contemporary problem of legitimacy arises from a disjunction between two bases for making collectively relevant political decisions. One basis is the economic and political bargaining mechanism of the market, which is supposed to mobilize individual calculations of interests into collective economic and political outcomes. The other, a tradition stretching from Condorcet, Turgot, and Henri de Saint-Simon to the present, is the technocratic "expertise" basis for decisions. Such expertise is allegedly called for by "objective" crises of relevance to the whole society that appear to defy solution by a market mechanism supposedly guided by purely "subjective" interests.

The American public has long accepted certain categories of problems, like physical illness, under the technical criterion but has resisted its application to other areas of life (Fine, 1956). A major change was heralded when the public came to accept the management of the collective economy as a quasitechnical problem after the Great Depression. Since then the public has been asked repeatedly to accept the transfer of many issues from the subjective category, amenable to market solutions, to the status of "objective public interest." This transfer implies new decisionmaking procedures and new technocratic elites to make the decisions. Issues allegedly requiring expertise are now coming to include the management of continuing international and domestic tensions (policy sciences); the resolution of distributive injustice rooted in accidents of social location and birth (welfare planning); and the control of population growth with its disruptive pressures upon all existing institutions. The struggle between proponents of these two orientations toward legitimacy of decisionmaking processes—consumer sovereignty and technocratic expertise—has itself clearly taken on the status of a major moral crisis of modern consciousness.

Pessimists expect this conflict of legitimacy doctrines to be resolved upon the emergence of a technicist ideology of societal management in the benevolent guise of social accounting, collective data banks, policy sciences, and cybernetic theories of immanent control mechanisms. This technicist ideology, stressing not social planning by real people as much as a myth of self-regulating (servomechanistic) cybernetic controls, is identified by the pessimists with sinister images of universal surveillance and administrative regulation of private life. In short, what is feared is the reconstitution of society from the nominalistic abstraction of the market into an objective whole whose real form will be that of a universal concentration camp (Gross, 1970). That this pessimistic imagery is shared by some pure economic liberals (for example, Friedrich Hayek and Ludwig von Mises, who reject the reification they see in collectivist implications of the word "economy"), by romantic Marxists (for example, Herbert Marcuse), and by some avowed theologians (for example, Jacques Ellul) serves to show again how irrelevant the left-right conflict is to an understanding of such pessimism.

It should be recognized that to analyze pessimist prophecy in this way does not necessarily mean opposing the humanist goals of cybernetic social planning. These goals, after all, have to do with establishing criteria for effective policies intended to increase human welfare. Indeed, the cybernetic metaphor is so seductive precisely because it directly addresses three such criteria; namely, relevance, accountability, and minimization of undesirable unanticipated consequences of purposive social action. Relevance and accountability are implied by the strong emphasis on input-output analysis, which is a feature of the cybernetic approach. Reduction of undesirable unanticipated consequences of action is implied in the notion, central to cybernetic thought, of feedback loops for system self-correction. Presupposed here, of course, is accurate and abundant information about what goes on in a system, information that can be obtained only through massive societal data gathering and storage procedures.

Pessimists are concerned about the possible totalitarian consequences of a centralized information policy and, more subtly in some cases, with the cultural effects of data categories relevant to the establishment of fixed norms for computerized decisionmaking. The possibility is envisioned that the cultural fixation of such categories will limit the range of concepts and possibilities that enter the realm of public policy discourse. If anything, this anxiety is stregthened by another seductive quality of cybernetic thinking;

its apparent ability to subsume and synthesize the assumptions of two master metaphors of Western social thought, organism and mechanism. The problem arises from a characteristic that cybernetic thinking shares with these metaphorical styles—reductionism. Unless heavily qualified, or leaded with anachronistic interpretations, the vocabularies of organism, mechanism, and cybernetics are not adequate for describing specifically human phenomena of freely willed projects, of intentional action, of *interpersonal* relationships.

Instead, reductionist metaphors—and this is a major reason for their popularity—conceptualize human behavior and society itself as determinate objects capable of controlled observation, measurement, and (in theory) predictability. Yet from a humanistic point of view this reductionist agenda, important as it is for scientific progress, requires persistent attention to a deceptively fine line between two quite different questions. The first is: *are* human beings and their works describable in terms of the assumptions underlying such metaphors? The second question is: is the history of modern societies characterized by social processes and cultural assumptions that operate to *produce* human beings who are constrained to act in ways that make it appear *as if* they were so describable? It is surely fair to say that indifference to the line between these two questions is an important element of a technicist mentality.

The idea that there may be structures of constraint that can produce people who think and act in ways that make them amenable to technological manipulation raises important questions about the relevance of constitutionally guaranteed liberal freedoms. Allen Schick, for instance, understands this well when he speaks of the possibility that

> in a full-blown cybernetic state, politics and bureaucracy would wither away, though their forms might remain. That is, there still might be contests for public office, but the process would not have its old importance. To the extent that sociostatic norms limit conflict, the scope of politics would be narrowed. Whether or not we ever reach the "political fiction" world of genetic or thought control, there will be less disagreement in the future than existed in the past. Already, in the macroeconomic sphere where the cybernetic condition is most advanced, differences between Democrats and Republicans now are minimal, despite the great stakes involved and the history of public controversy (1970:12).

Here again we have the question of where the locus of effective decision-making is to lie: "Of what use will be the right to speak if the speaker has little ability to challenge the dominance of the experts and the systems engineers and little ability to sway the course of public policy" (ibid.:23). I shall return to these issues in greater detail in Chapter 3. In this chapter they have been noted only in connection with the growing tension between two models of self-regulating systems: the market economy and the cybernetic society.

COMMUNAL AND ASSOCIATIONAL ORGANIZATION

This distinction has to do with the various established assumptions about what it is that human beings in a given society have in common, owe each other, have the right to expect from each other, how they identify with each other. Sociologically, the distinction points to different modes of social integration and cohesion. Popularized by Ferdinand Toennies (1957) in a book on the concepts *Gemeinschaft* and *Gesellschaft* published in 1887, the distinction is all too often used as if it were a rigid dichotomy. Also, the concept of *Gemeinschaft* has been associated with a kind of romantic primordial unity of absolute identification. [13]

The community-association distinction is best utilized without many assumptions concerning the degree of actual psychic identification with these modes of organization, on the part of participants. One should also not confuse community and association with any rigid views about the size of groups involved, such as might be suggested by the overlapping distinction in sociology between primary and secondary social relationships. Although community obviously implies a greater degree of primary intimacy than does the concept of association, such a one-to-one correlation of these distinctions would ignore the fact that communal enclaves occur within associational contexts, and that all communities have associational elements.

What the community-association distinction is really pointing up are two abstract principles of moral identification and integration, principles located not necessarily in each individual's heart but in the society's world view. This and related typologies [14] were often used under the assumption that historical evolution is a progression from one end of the polarity to the other. The status of such evolutionary hypotheses is now a matter of considerable controversy, having been subjected to criticism on grounds of faulty methodology and ethnocentrism.

In our context we speak of the community-association distinction simply in terms of two moral approaches to how people's interests might be defined. It may be assumed, in general and in principle, that the ultimate good that should inform the substantive purpose of a person's life is something that is known prior to and outside that given individual's existence. The common social good then has to do with facilitating each person's access to a commonly valued way of life. This orientation I would like to regard as defining a communal principle of cohesion. In the context of such an orientation, people have more in common with each other than not. Doctrines of liberty, if they exist, will tend toward positive justification; that is, liberty to achieve some defined goal or state of being. Whatever the conflicts of medieval life, for example, and they were many and violent, it was more or less taken for granted that everyone had an immortal soul that needed salvation. Therefore, the conditions of salvation were not matters left to random private judgment.

The purely associational principle of identification, in contrast, is based on a much more limited, utilitarian notion. It presupposes an extreme moral relativism of values to a degree in which the only credible basis for legitimate public decisions is common denominator utilities. Liberty is justified negatively; that is, liberty to define one's own ends since, as Bentham said, "there is no one who knows what is for your interest so well as yourself."

Again, therefore, it should come as no surprise that the destruction of the religious unity of the Middle Ages created a generic moral problem of human affiliation and consciousness-of-kind that in the nineteenth century entered repeatedly into sociologies of classical theorists. From Henri de Saint-Simon to Karl Mannheim, sociologists tried to resolve the problems created by relativism. They understood that they were witnessing the fragmentation of modern civilization into powerful centrifugal forces. These forces consisted of the random, interest-conscious, human egos of political and economic liberalism; the class-conscious strata and occupational subcultures generated by the technological division of labor; and the polities increasingly dominated by nationalisms. Such thinkers understood that relativistic and centrifugal tendencies of this sort were capable of undermining all established sensibilities of common human *solidarity* (Hayward, 1960).

At issue for social theorists was the increasingly ambiguous relationship between subjective freedom dominated by personal calculations of interest and valid criteria of obligation to collective social, cultural, and political units. This concern for defining the proper bases of moral and political obligations (and for legitimate coercion) has persisted against a backdrop of increasing institutional integration in modern civilization. This general problem of legitimate obligation is aggravated by a growing inability of local administrative entities to respond spontaneously and effectively to the problems of production, distribution, security, and ecological imbalance generated by industrial processes.

In the face of these latter pressing trends, the pessimist argument builds a dark prognosis that is sometimes traced back to the resolution of the problem of order and solidarity first proposed by Saint-Simon. With him utility achieves its highest status as a standard for the architectonic reorganization of whole societies. He had discovered the one unquestionable collective utility for all men: the principle of productivity itself.

> The effect of the passion for equality (wrote Saint-Simon in 1817) has been to destroy the social organization which existed at the moment of the explosion. I therefore ask whether, once this has been destroyed, another notion is not needed to inspire the work of creating a new organization. . . . Where shall we find ideas which can provide this necessary and organic social bond? In the idea of industry; only there shall we find our safety and the end of revolution. . . . The sole aim of our thoughts and our exertions must be the kind of organization most favorable to industry . . . including every kind of useful activity, theoretical as well as practical, intellectual as well as manual (1964:69–70).

The anxiety that this technocratic tradition has generated is based upon the totalitarian imagery of order and integration suggested by the program of coordinating culture and society around what is, in essence, a purely technological goal (Iggers, 1958; Talmon, 1960:35–124). The contemporary version of this anxiety can be expressed in terms of two predicted, interlocking developments that are seen as pointing to a technicist order. The first development is the massive spread of functional organizations (occupational, professional, and administrative) as settings for membership groupings in place of the family, the voluntary association, the church, and the integral territorial-political community. Indeed, this was the program espoused deliberately by Emile Durkheim as the solution to the problem of anomie that he had done so much to formulate (1958:208–220, 1964:1–32). But his intention was to reconstitute, in service of the fight against anomie, the medieval principle of corporate obligation, security, and action. Durkheim did not see that modern work processes were going to be too different from the medieval mode of production to allow such a solution—perhaps because he was too desperately seeking a solution to the problem of anomie, which tormented him so much.

As a result of these trends, the second development the pessimists envision is the confrontation of the essentially helpless, atomic individual with what will come to be the total power of large, interlocking state and corporate organizations. It is feared that this fusion of state and society will replace the public identity of citizenship with a set of roles in which the person functions, be it knowingly or not, merely as instrument, a cybernetic operating unit (Boguslaw, 1965:99–126). The ends for which such an individual exists would allegedly come to be formulated neither in terms of common religious, aesthetic, or ideological traditions nor in any sense by the person as an independent moral agent. Instead, the ends dominating the individual's public life in a technicist society would be (1) *defined* in general by the logic of the technical processes of production and societal servomechanistic regulation; (2) *designated* for any given situation by systems-engineering elites through the medium of roles decreed as socially necessary (the moral basis for the "job morality"); (3) relatively *unaffected* by other values; and (4) *legitimated* by common denominator utility rewards like security, status imagery, money, and invitations to feel pride in the furtherance of abstract technicist myths like the gross national product.

Whether some new "global village" of the middle classes would ever be capable of inducing communal reflexes and sacrifices on cue is, of course, another question. In any direct sense the answer is probably no. But then few natural communities, so to speak, are normally capable of such integration. In the frankly totalitarian societies it is quite likely that dependence upon the media of communication is more resisted by blase and cynical populations than it is in democratic societies, where the constraints on public communications are so much more subtle and informal. Accordingly, social critics like Jacques Ellul (1965) and Ivan Illich (1970) have put forward the hypothesis that the most politically significant function of public education now has become that of preparing the fledgling average man for appropriate symbol literacy in the global village. Indeed, for Illich this interpretation justifies total resistance to the establishment of schools in developing countries.

For the pessimists such considerations abort confidence in any facile solutions (like increased leisure or education) for alleviation of the posited trends toward technicist coordination. What, then, is one to make of the libertarian hopes so salient in the literature of participatory democracy on both the left and right sides of the political spectrum? Are such expectations without any reasonable foundation? If not, what alterations in the communal and associational aspects of social organization could accommodate the notion of a libertarian technicist society?

Although an apparent contradiction in terms, a libertarian technicist society seems possible in principle. It would require an intensified privatization of moral norms and values, combined with an obsolescence of the principle of privacy itself. Let us see what this means.

LIBERTARIAN TECHNICISM

The technicization of culture has not gone unresisted. One important popular source of resistance has been the so-called counterculture for whose spokesmen technicism and totalitarianism are virtual synonyms. This view is based on a fear that *communitas* may disappear, swallowed up by *societas* in a vast administrative concentration camp. This is not necessarily a reasonable forecast. The psychic basis of communitas is intimacy; of societas, impersonality. There is little evidence that people can long do without the communal bonds of parochial identifications. Intimacy, it is safe to say, is a permanent need of the human heart. Bureaucratic impersonality facilities broad functional interdependence and allows the assimilation of complex technologies into society. However, it contradicts the intrinsic values of intimacy, loyalty, and local narrative mediated by communitas. The question, then, is: how are societas and communitas to be reconciled when society is expanding beyond intelligibility from the standpoint of the person as citizen?

Certainly the ideologies and utopias of communitas are not dying. Under the banner of personal and group libertarianism, the exponents of voluntary community are seceding, at least emotionally, from the ranks of what they already think of as the technocracy. Yet few of the spiritual nomads encamped outside the walls of Technopolis appear ready to wander too far from the marvels of the city into the truly primordial forests. We will, I think, witness the gradual emergence of a pattern of existence that combines functional residence within the walls and emotional residence without. This prophecy may be labeled the "libertarian technicist" society. It rests on a near total abandonment of the effort to reconcile communitas and societas in any superordinate moral form whatever except minimal common utility. At the same time, this would represent the most extreme, viable intensification of the present version of the distinction between "public" and "private" domains of authority.

Relative to present liberal market societies, a libertarian technicist society would be characterized by a combination of a heightened technological rationality in public life and a nearly total privatization of the community principle. The society at large would have almost none of the properties of a moral community. Instead, the public order would be envisioned as an impersonal system of gyroscopic controls.

In such a society, breaking the law would not be a moral violation but a technical one—a disturbance of cybernetic equilibrium. Every effort would be made to keep statutory norms to a minimum, justified not by moral doctrines of the good but by cybernetic rationality. Punishment would be secondary to the principle of system self-regulation, as embodied in mechanisms like punitive taxation and other appeals to utilitarian common sense. Many definitions of deviance now subject to public legal formulation and sanctions would be transferred, in such a society, from the public level to the voluntaristic communities of the public realm. For example, society at large would probably become indifferent to most forms of drug taking, exotic sexual behavior, and even to some varieties of what are now called mental illness.[15]

This would likewise happen with respect to positive evaluations of various kinds. Thus, the state would have divested itself, in such a society, of the right to set standards of quality in virtually all areas of certification except those relevant to societywide roles of common utilitarian importance like medicine or activities in which direct state financing allowed the piper to call the tune. (Some activities now directly financed by state revenue would come under the voucher payment principle.) Rather than setting standards, the state would become just one more consumer of diverse talents on the open skill market. Like other institutions, the state might have its own training facilities, but these would be no more "official" than those that could be set up by any other institutional sector like industry, the church, or private research foundations. The state would reserve for itself a monopoly over facilities such as medical schools and public policy science institutes. But all this would be more in the name of insuring technologically rational job specifications in public utilities than of perpetuating any moral principles other than efficient utility itself.

A state minimum-income security welfare policy would exist in such a society, kept within bounds by strongly enforced birthrate policies. Since the emphasis on work would itself have come to be defined as a relativistic moral value, the state would have to downgrade its role as the moral overseer of the Protestant work ethic. Instead, a combination of free public services and minimum-income floors would provide for general security according to a formula derived from a mixture of social accounting rationales and political processes. It is likely, too, that population rates would be governed by cybernetic considerations, the state being allowed to issue tradable child-bearing licenses according to some flexible demographic formula. The term "equality of opportunity" would expand to take on a libertarian overtone. It would include the right to life-styles from near idleness to work in the service of a wide range of values sanctioned by a spectrum of privatized communities.

For a libertarian technicist system to operate, by definition two requirements must be balanced. One is optimal latitude for private freedoms. The other is providing the state sufficient information about aggregate social processes to avoid the instabilities that lead to disorder or to (cybernetic) irrationality. The balancing of these two considerations implies what to liberal ears is still a contradiction in terms, but in a technicist society would not be. The only way that a "do-your-own-thing" standard of private freedoms can be reconciled with the generating of enough information to manage a stable public order with the aid of computers is the sacrifice of privacy. There is only one moral justification for sacrificing privacy that is compatible with a libertarian philosophy. That justification is the reduction of information to the symbolic status of morally neutral "data" for exclusive use by the social planners and their computers.

This is conceivable as a justification because it is part of the meaning of a technicist society that the state has become morally neutral to information, its interest directed purely to information's significance for cybernetic regulation. In such a society law enforcement officials could not count on the moral indignation of the public. A law violation being an act for complex technical definition and not a matter for common moral comprehension and censure, the police would be reduced in the eyes of the general public to the abstract status of employees of the state system engineers.

In our time, when political discourse is being reduced to either battles between putative experts or expressions of universal suspicion, a libertarian technicist model of society may come to seem the only one in which the tensions between societas and communitas are theoretically reconcilable short of totalitarianism. Since the demarcation lines between societas and communitas are ever fluid (subject to shifting resources, changing conditions, and varying definitions of what needs regulation), the substance of politics in such a society would be the ongoing negotiated trade-offs between agents of the cybernetic state and partisans of the various privatized communities.

This, then, is the theoretical outline of a libertarian technicist society—a do-your-own-thing system. The latitude of such freedoms would depend primarily upon two factors. One is the level of automated production existing in the society; this affects the amount of free time from labor discipline that can reasonably be afforded. The other factor is the international situation, which determines the safety status of the society as a whole and hence the level of internal mobilization for conflict, both material and psychic, that must be maintained.

Inevitably there is what for some would be a heavy price. A libertarian technicist society would presuppose a permanent schizoid dissociation between two kinds of reality. One kind is expressible in the public, mathematical, and therefore minimally intelligible but universally legitimate language of official positivism. The other kind of reality is the relativistic domain of multiple, aesthetically rich, and morally directed private worlds. The psychic rewards in such a society would go to personalities that were flexible, unserious, not characterized by drives toward integration of social and private selves, and able to withstand a kaleidoscope of relativism by reducing most of it to play. Serious departures from this psychic ambience could result in perceiving such a world as insane.

A libertarian technicist society would not be characterized by any collective eschatological motifs common to present Western social philosophic traditions. Whatever the inconsistencies and conflicts within and between societies of liberal and Marxist cast, they are—however indirectly—all dominated by a mythologized ethos. That ethos is the notion of collective moral progress, of hope: hope at least for men's continued sense of common human venture; hope at most for the perfectibility of persons according to some vision of civilized excellence.

Consider Saint-Simon's ideal of cooperative rationality; John Stuart Mill's desire for the spread of the "higher pleasures" of civilization; Marx's ideal of surmounting the dualist tensions of "individual versus society" and "theory versus practice"; Durkheim's expectation of organic solidarity; Weber's wish for the renewability of charismatic inspiration; or Mannheim's planned freedom. We find in these and other themes of classical social theory persistent hope for progressive redemption from the present. The promise of a technicist society, however, is not civilizational progress but societal stability. Its credo is adaptation and adjustment to its own system properties and to those of its "environment"; its destiny, an age of play at the end of time. Technicism is a public world of stasis, interrupted from time to time by some great technological breakthrough in epic adventure (for example, space travel) or provision of a new common utility (for example, cancer cure). Only in its interstices would be found the prophets of progress and integral civilization, awaiting the dispensation of some new revelation through which men could once more discover for a time their *common* destiny.

The Metaphysics of Social Diagnosis

We have seen that few contemporary antitechnicist writers have had the intention of centering their criticism exclusively upon the aesthetic defacements or other direct impacts of physical technology. Although certainly aware of these effects, most critics are neither pastoralists nor primitivists.

We have also seen that, on the whole, it is not being argued that elite conspiratorial motives, such as conscious desires to protect vested interests, is the key problem. One finds few references to throwing the rascals out and bringing in the virtuous. Nor is amelioration to be achieved simply through procedural reforms such as new participatory decisionmaking mechanisms. Rather, what is at stake is a profound ambiguity in the cultural definition of what participation itself is to mean in an epoch of great technological power. This sort of problem is not just another moral issue among many. This is why technicism represents for its critics a dehumanization, a reduction of potency, so to speak, in some essential way of being human. Most fundamentally, this dehumanization is seen as a diminishment of opportunities for the exercise of personhood.

This chapter has stressed social organizational themes. But the problem of participation stretches, along the spectrum of humanist criticism, from the question of what contemporary people are allowed to *experience* of the world all the way to the question of what they are allowed to *make* of it. These two issues, that of the available structures of experience and the permissible range of action, form the axes of perhaps the single greatest moral theme of Western civilization since Athens and Jerusalem: the spiritual itinerary of the self as moral agent.

Beyond this general concern for the dignity of the self, critical perspectives vary greatly. For the sake of further clarification I would like to venture a simple and certainly somewhat misleading classification. It would be useful to understand the variety of metaphysical loci, or horizons of interpretation, of antitechnicist criticism. The benefits are obvious. To understand even a few of a thinker's basic assumptions is to have some insight into how he looks out upon the world, what his priorities of questions are, what he worries about, what he hopes for. Seriously undertaken, this is obviously the task for individual biographies. On the other hand, with the aid of but a crude and provisional classification, it is possible at least to begin to understand how it is that a group of thinkers representing divergent disciplines, ideologies, and nationalities can come to have a common pessimistic vision of the future. To avoid constructing procrustean beds, there will be no effort to force particular writers into exclusive categories. Some have explored myriad implications of a single metaphysical perspective and others have not. There seem to be three metaphysical standpoints from which diagnoses of technicism have been forthcoming. I shall call them *egoistic rationalism, neoromantic humanism, and gnosticism.*

If one begins with the view that it is always intellectually illegitimate to define social reality as anything other than people calculating their utilitarian interests, the whole terminology of collective socioeconomic concepts will tend to be regarded as incipiently technicist. Thus, anyone taking upon himself the right to make decisions affecting other people on the ground of alleged insight into the "laws" of "history," "society," "economy," and so forth is considered, in this view, as asserting illegitimate, pseudotechnical knowledge. Most markedly, Friedrich Hayek and Ludwig von Mises have been associated with this perspective in their antitechnicist writings.

Neoromantic humanism is a catchall phrase that encompasses antitechnicist criticism originating in notions about the capacities and needs of man that are considered peculiarly vulnerable to deformation by technology and its various effects. The phrase covers criticism from pastoralist, primitivist, medievalist, and neoMarxist standpoints. Despite this variety, the term neoromantic is used here because of a characteristic that such styles have in common. Validly or invalidly, it is a rather exalted view of the uniquely human characteristics necessary to happiness that are seen as threatened by technological rationality and materialism. Sometimes these attributes are considered as having been achieved and now lost. For others it is the potentiality for further humanization that is at stake.

Thus, if the reference point is the past, nostalgia is often the emotional aura in which the past appears. The politicization of such nostalgia is an important theme of the study by Fritz Stern (1965) of pre-Nazi antimodernist ideologies in German culture. Likewise, Benjamin Schwartz considers Hannah Arendt's "religion of policitcs" to be in the tradition of those German intellectuals at the end of the eighteenth century among whom although "alienated from all actual political experience . . . we find the most exalted dream of the life of the ancient polis" (1970:148). If the reference point is a utopian future, we similarly find hope for human capacities largely ungrounded in historical experience. As Martin Jay has said of Herbert Marcuse, "It is only as a utopian hope that the coordination of self-creating action and rational theory should be understood in Marcuse's work" (1970: 344).

Lewis Mumford is perhaps the most important current exemplar of a less apocalyptic Anglo-American tradition of antitechnicist romanticism. Despite criticisms leveled against him for overgeneralizing the influence of science upon technicist trends, Mumford presents in his various works the most detailed analyses, perhaps since William Morris, of what is meant by the charge that technological civilization destroys the individual's capacity to take part in the craft of fabricating his world.

Finally, there is what might be called the viewpoint of modern gnosticism. The reference to gnosis here addresses its most general sense: a conviction that there is "an esoteric knowledge of higher religious and philosophical truths" that can be attained only by an "effort to transcend rational, logical thought processes by means of intuition" (Runes, 1959; pg. 117).

In this perspective, of which Heidegger is the exemplar without peer, technology is raised to the status of a world view. Here science is itself considered to be part of the ontological history of technology as are not only all modern institutions but, indeed, the entire enterprise of Western metaphysics since Plato. The knowledge considered hidden by technology—the latter understood as perceiving and organizing the world in such modes as are necessary to having power over it—is true being. The logical definition of being as a determinate object is, for Heidegger, a contradiction in terms.

Perhaps more controversially, a case can be made for including Jacques Ellul under this designation. His popular books on the concept of technological society (1964) and on propaganda (1965) appear as almost exclusively sociological analyses. As such, Ellul often seems bafflingly overdeterministic in his depiction of the all-pervasiveness of technique as the modern myth. However, to read his lesser known theological works (1960, 1967a) is to recognize that Ellul's standpoint is that of a Christian who experiences modern society as closed off, by its concern with the technicalities of the finite, to grace and transfiguration by infinite divinity. Although eschewing synthesizing systematizations, Ellul has himself commented that "the reply to each of my sociological analyses is found implicitly in a corresponding theological book, and inversely, my theology is fed by socio-political experience" (Holloway, 1970:6).

Summary and Conclusions

My intention in this chapter has been to synthesize the literature of what could be called technicist futurology into a statement of social and cultural theory. Two limitations are immediately evident. One is that the discussion has been confined to the dynamics of liberal societies. This was deliberate for reasons of space and complexity of topic. Another reason for this limitation was the convenience deriving from the fact that of all ideological systems, liberalism seems on face value most antithetical to the prophesied features of technicism. The second evident limitation is that my synthesis is incomplete, a lack clear to anyone familiar with the literature of criticism that has served as a baseline. The horizons of interpretation from which such literature derives are simply too varied and contradictory for facile reconciliation on any single level. Yet I have tried to cover enough of a range to persuade the reader of two points. One point is that the pessimist critique does have a coherence of its own and is not just a series of dyspeptic reactions to diverse unpleasantries of modern society. Second, the critique is not an emanation of one single political ideology.

The findings thus far may be summarized by way of five general conclusions.

1. Among the ways in which it is profitable to seek an understanding of liberalism, one can regard it as a set of institutions that provides a particular response to a cultural transformation of world view in the West. This new world view is one characterized by philosophical materialism, determinism, and value agnosticism (if not nihilism). Seen in this way, the major latent function of liberal ideology and practices is to protect two long-standing ideals of Western civilization against the fullest implications of the new secular world view. These ideals are the dignity of a uniquely human status in nature and the notion of personal authenticity, symbolized in the doctrine of free will.

Liberalism as a whole achieves this agenda at a price. One element of this price is that liberals have been forced to respond to agnosticism, materialism, and relativism by reducing collective vocabularies of evaluation to calculations of utility and by reducing the person to little more than a consumer of economic products and political opinions. However crass it may appear to the aristocratic mind, utilitarianism has served as a buffer against both capitalist autocracy and the sorts of ideological mystifications that have transformed the Marxist vision of freedom into a totalitarian nightmare. Liberals could achieve this, however, only by taking for granted the assumptions of dualism forced upon them by the scientific-materialist world view. Thus, liberal societies are organized around polarities like individual versus society, theory versus practice, and subjective versus objective. The most influential criticisms of liberalism have been inspired by the hope that such dualisms could somehow be surmounted.

2. The demonic vision of a fully technicist society, as far as its connections with liberalism are concerned, is really based on one implicit, general proposition; that is, that liberalism's ability to preserve the claim of uniquely human status and the ideals of free will against erosion by cultural modernism is an unstable phenomenon. It is unstable because a social system supposedly cannot indefinitely maintain itself against the *major* implications of its philosophical grounding; that is, the dominant assumptions about the nature of reality that its most critical scientific intellects take for granted. Thus, *technicism is really the intensification of the modernist world view in the area of its implications for social structure.* Like the slow self-correction of a geological fault, technicism reflects the gradual alignment of modern society with modern scientific culture.

3. The specific mechanism generating the changes leading to this alignment are crises of mass welfare and, potentially, even survival. The logic of this argument in the pessimist perspective is as follows:

Technicist trends are legitimated by appeals to collective human welfare at the expense of agendas defined generally as the development of personal authenticity. The value agnosticism and philistinism of the modern world view engenders a false distinction in liberalism between public and personal interests. This is because public interests are understood as the welfare of collective units, and personal interests are confused with private, hence random, relativistic and individual interests. Under circumstances considered normal (that is, crisis-free), and subject to a few obvious limitations, private interests are considered morally equal in their claim upon public toleration.

In a liberal society it is ideologically very difficult to institutionalize a *public* cumulative heritage for *personal* development (witness the irrelevance, as mass ideals, of concepts like "cultivation," "noblesse oblige," and "civilized excellence"). In times of crisis, any conflict between public interests and private interests tends to be defined as a tension between "social responsibility" and "shortsighted vested interests." This is an outcome of the dualistic antipodes "individual" and "society" in periods during which the collective welfare seems threatened. Likewise, when resistance occurs against such reification of the social, it does so increasingly under the banner of an almost solipsistic philosophical ambience (do-your-own-thing).[16]

To put it more directly, it is easier in a liberal society to achieve agreement and collective action on what is necessary shelter as against what is beautiful housing, on jobs as against work, on income as against progress, on quantities as against qualities.

4. The result of what has just been described is that each publicly perceived crisis will intensify tendencies toward coordination (not necessarily centralization) of decisionmaking and toward more efficient social control. In the absence of any other bases, the criteria for order and efficiency will be common denominator utilities. These are power, security, and predictability of changes with reference to the status quo.

Since the meaning of the status quo is not necessarily itself a matter of common agreement, such definitions will themselves be propagandized by the major contenders for control of state and corporate resources. These conflicts between social and political partisans will encourage the legitimation of a positivistic vocabulary of "policy science." Policy science proponents will professionalize themselves and attempt to gain control over the command posts of society on the ground that they are morally neutral as regards all political partisans, being concerned only with the welfare of the "whole." Once in control of advisory posts, policy scientists need not occupy political offices to wield power because they will have become guardians of public policy *assumptions* and their language.

5. Technicism is a self-fulfilling prophecy to the extent that scientific and technological vocabularies are used by their partisans deliberately to downgrade and displace vocabularies of free-willed action in the interest of solving concrete social problems. No criteria for public agreement on no-

tions of personal authenticity having been achieved, these latter considerations are vulnerable to being dismissed as too abstract. Under such circumstances the partisans of ideals of personal authenticity are deprived of publicly legitimate theory and hence are increasingly tempted to turn to apocalyptic and anti-intellectual orientations of spontaneous (theoryless) action. The motto for such pure praxis could well be the phrase that appeared on the walls of Nanterre University in 1968, "All power to the imagination." The subversive and totalitarian possibilities of spontaneous praxis can be expected to generate fears that intensify public support for technicist controls.

It is worth stressing that the type of analysis in which I have engaged is quite different from the extreme deterministic demonologies that wind up in downgrading technology as a cause of dehumanization. For one thing, technicism as described in this book is nowhere as yet a social fact and in a certain sense perhaps can never be. Nazi Germany is often posited as a paradigm case of technicism. The British historian Geoffrey Barraclough (1971) has leveled heavy criticism against this view. Quite aside from presenting many examples of technologically irrational standards of action characteristic of that regime, Barraclough gives telling examples of the argument that reactionary antitechnicist rejection of modernism was a prime theme of German culture inherited both by Nazism *and* by some partisans of the most dedicated anti-Nazi resistance alike. He delivers a much needed warning in his observation that the tendency to view Nazism merely as a symptom of a social order already stained with technicist original sin actually deprived the resistance of a concrete program. The resistance appeared hopelessly abstract and even self-destructively pessimistic in relation to the seeming dynamism of the Nazis.

Furthermore, to consider technicism an actual or eventually inevitable fact is really to accept a completely deterministic metaphysics. From the standpoint of any voluntaristic approach to social theory, technicism is a tendency, perhaps a trend. It is possible to argue that for a variety of reasons people actually desire what has been defined as technicism. This can be because people find it easier to accept technicism than to pay the price of reversing it, or because people want its positive benefits, or because whole populations have become sufficiently illiterate as regards *nontechnicist* sensibilities and styles of thought that effective resistance no longer exists.

It should be clear also that it would be a most dangerous course of action for social critics to allow themselves to oppose technology, planning, and cybernetic techniques wholesale in the name of opposing technicism. That would again be doing what one fears—reifying tools into causes. A more fully technicist future is a possibility, but if it comes its victory will be based on two factors. One is the capacity of such a society to solve certain concrete problems requiring rational planning and societal coordination. The other is a failure on the part of populations to have achieved the technological and nontechnological literacy necessary for effective voluntaristic public action, literacy that today must include the capacity to cooperate in controlling complex social and physical tools to insure human survival.

But is there yet a conception of human dignity that can underwrite the ideals of this literacy and act as a critical brake upon the advance of technicist culture? To this question we must now turn.

Notes

1. These examples are easier to cite than to establish empirically. As Converse (1964) has made clear, very little knowledge may be safely assumed about the social distribution of values, information, orientations toward action, and states of awareness generally among a large modern population. Mann (1970) has reopened to critical analysis the assumption that legitimation through value consensus is a necessary base for social order. He has not disproved, and really did not attempt to do so, this proposition but rather to raise some searching questions. Among those suggested by his review of relevant empirical studies are the degree and distribution of value consensus and dissension. From a historical perspective, however, it would be difficult to deny the importance of value consensus on some level. A society organized around the "commonsense" assumption that a man's purpose in life had better have some relationship to the salvation of his immortal soul simply has to be different from one in which it is more or less taken for granted by large masses of people that "one doesn't go around more than once." Of course, this not to suggest value *determinism*. That would be a very different kind of argument.

2. The example of medicine is cited from Cassirer (1964:142). Other valuable studies in the cultural history of problems surrounding the definition of man in nature for the early modern period and the period just preceding it include Haydn (1950), Huizinga (1954), Hazard (1963), and Willey (1953). For the problem of usury see Nelson (1969). For studies having to do more specifically with the transformation of views of nature, especially as regards the theory of motion, see Dijksterhuis (1961) and Thomas S. Kuhn (1957). Particularly helpful studies of the impact of these changes, especially in theories of motion, upon social affairs and religious-moral perspectives are Koyré (1957), DeSantillana (1955), and Prosch (1966). An outstanding examination, for both its substance and literary style, of the impact of the new theory of motion upon Hobbes and the whole subsequent mainstream of political theory down to the present day is Spragens (1973). A valuable study of the religious foundations in the sixteenth century of the later secular conviction that human history is a source of insight about human nature (an important assumption of social science) is Rudnick (1963).

Obviously, such studies are valuable for the light they cast upon the changing notions of what it means for human beings to "participate" in nature and society. They also show that the changes we discussed in this study did not occur all at once. Nature was being looked to for moral guidance in one way or another as late as the early twentieth century. My text selects out of the complex and tumultuous culture history of these centuries certain tendencies for highlighting because these tendencies are the major constituents of the pessimist argument about technicism. For any other purpose the brevity of this description would do a disservice to an under-standing of the issues and passions involved in the story. It should also be added that many of the themes of modernism such as the importance of personality, nom-inalism, and social contract can be traced well back into medieval times. I do not in-tend, by contrasting modern and premodern times, to imply a sudden, total trans-formation. Rather, I stress certain key changes whose cumulative impact we now see in hindsight eventually added up to a new world view.

3. It should not be thought that these tendencies remained unchallenged. Apart from Hegel's great effort to reconcile the subjective and objective realms of being once again, the countertrend remained strong on the level of social theory, too. For a view of nineteenth-century sociology as itself influenced by these con-servative tendencies whose proponents sought to find ways to reconcile the subjec-tive individual with social organization, customary traditions, and political authority see Nisbet (1966).

4. My use of the term "nihilism" has been influenced, aside from Nietzsche and Heidegger, by Jonas (1958), Voegelin (1952), Loewith (1964), Kahler (1957, 1967), Rosen (1969), Thielicke (1969), and Barrett (1972).

5. The most relentless and detailed philosophical exposition of this argument is the works of Heidegger. On the sociological level of the issue, Weber probably re-mains without peer. A more limited but appropriate introduction to the sociological form of this assertion is Polanyi's *Great Transformation* (1944).

6. For a provocative statement of the moral and social implications of this shift see Gouldner (1970:61–88).

7. This last set is less immediately recognizable. It really refers to the distinc-tion better known to social theorists by way of the German terms *Gemeinschaft* and *Gesellschaft* thanks to Toennies's popularization (1957). They are generally translated as "community" and "society," respectively, but I refrained from this usage because of the confusion that might result from the different sense of the term "society" in two of the three sets of phrases we are examining. Accordingly, I adopted the terminology utilized in Chapters 1 and 17 of MacIver and Page (1959).

8. Given the large body of literature relevant to this point by aristocratic and bourgeois critics over many generations, the question naturally arises, for sociol-ogists, why it has had so little popular effect. For this, another kind of investigation is necessary: the study of what the French call *la vie quotidienne*; the phenom-enological (not ethnographic) study of "everyday life" in terms of its recurrent pat-terns. Lefebvre, in his controversial book on the subject, puts it this way:

> The quotidian is what is humble and solid, what is taken for granted and that of which all the parts follow each other in such a regular, unvarying succession that those concerned have no call to question their sequence; thus it is undated and (apparently) insignificant; though it occupies and preoccupies it is practically untellable, and it is the ethics underly-ing routine and the aesthetics of familiar settings. At this point it encounters the modern. This word stands for what is novel, brilliant, paradoxical and bears the imprint of technicality and worldliness; it is apparently daring and transitory, proclaims its initiative and is acclaimed for it. . . . In this [modern] world you just do not know where you stand; you are led astray by mirages when you try to connect a signifier to a signified— declama-tion, declaration or propaganda by which what you should believe or be is signified. . . . What can reason have to do with everyday life and modernity? What connection can there be between the rational and the irrational (1971:24–25)?

One of the important properties of everyday life, then, is the presence or absence of style and ritual. This is connected with the general degree of "world" integration among culture, society, and self. In a section on "terrorism and everyday life," Lefebvre argues that modern society

> no longer constitutes a system . . . but only a lot of sub-systems, a conjunction of pleonasms threatened with mutual destruction or suicide. Thus it is not really surprising if obsessional integration and specific limited integrations (of publicity to trade, programming to everyday life) lead to a sort of generalized racialism stemming from the disability to integrate properly: everybody against everybody else; women, children, teenagers, proletarians, foreigners are in turn subjected to ostracism and resentment, becoming targets of undefined terrorisms while the whole is still held together by the keystone of speech and the foundations of everyday life (ibid.:183).

Science fiction writers occasionally achieve in literature better depictions of this sort of analysis than do professional thinkers. A first-rate example of this is Brunner, *Stand on Zanzibar* (1969). This book reads as though it were written to illustrate the paragraph just quoted. A rather different (and largely non-Marxist) effort at a phenomenology of modern everyday life is Zijderveld, *The Abstract Society* (1970). Zijderveld concerns himself primarily with the phenomenology of a society experienced as an abstract entity, affording the self a subjective experience quite different from what we mean by being grounded in a "world." Also useful for those interested in such matters is the general phenomenology of "relevance structures" in everyday life developed by Schutz and Luckmann in *The Structures of the Life-World* (1973).

9. Parenthetically, it should be added that this generalization need not take the form of an officially institutionalized term or concept of truth, just as many societies obviously based on religious commitments did not have a specialized word for religion. Furthermore, there is no implication here that people in general had to be philosophers in their ability to justify the asserted notion of truth any more than the modern television viewer must understand his television set. It is only necessary, in both cases, to believe that *someone* knows the details of the truth. How many average persons of medieval times, for example, understood the theology of the sacraments, which formed for them the ritual axis of participation in the divine?

10. Probably the most poignant expression of this view of the legitimate function of politics in a liberal society, by a man who was clearly ambivalent about its implications, which he so well understood, is Weber's essay "Politics as a Vocation" (1958). For a moving statement about Weber's position in the general history of social thought see Nelson (1965c).

11. For arguments that alienation, though never used again as a word by Marx, was ever with him as a problem, see Lobkowicz (1967), Kamenka (1962), and Avineri (1968).

12. Aside from the various nineteenth-century masters of thought and sensibility associated with this theme (Kierkegaard, Nietzsche, Ruskin, and Morris), the issue appears in many guises in contemporary analyses such as those of Kahler (1957, 1967), Heller (1959, 1968), Graña (1964), Sypher (1968), and Berger et al. (1973).

13. Wrong (1966) has heavily criticized this form of the ideology of community. For a justification of my use of the terms "communal" and "associational" see note 7 above.

14. Status to contract, folk to urban, mechanical to organic solidarity, and particularistic to universalistic role ascription are a few other evolutionary typologies well known in sociology that are more or less loosely related to the *Gemeinschaft-Gesellschaft* distinction. One should not confuse these distinctions with each other since they reflect important and discrete bodies of work in their own right. Furthermore, they refer to different empirical points of reference. On the other hand, they all arose out of a common sense of striking differences between contemporary modern and all other kinds of society and thus represent attempts to come to terms with this conviction by establishing theoretico-empirical criteria.

15. Horowitz and Liebowitz (1968), and subsequently others, argue that the traditional distinction between social problems and the political system is becoming obsolete because what was once defined as social deviance is now becoming viewed as political marginality. This development is in line with the logic of a libertarian technicist order. It should also be noted that the concept of community in our analysis is not restricted by the criterion of isolated spatial location. Given modern technology, communities can be spread out over large spaces, interpenetrating with the organization of society itself. I have studied the history of the Jehovah's Witness movement from precisely this standpoint; its transition from a geographically separate community to a large-scale community dispersed within the interstices of its environing society (Stanley, 1969).

16. This generalization is evidenced by the current popular reduction of moral analysis away from ends and goals unto the level of virtually pure "process" for its own sake. This can be seen in the history of the "encounter group" movement and the anticurricular emphasis on spontaneous "learning." For paradigmatic criticisms of these movements from this standpoint see Maliver (1971) and McCracken (1970).

3 Species Survival and Human Dignity

REFLECTIONS ON THE MORAL PHILOSOPHY OF SOCIAL ORDER

A fundamental intellectual problem posed by pessimistic critics of modern civilization is the manner in which social order is to be legitimated in our problem-racked era. Order is meant here in the most inclusive sense: the evolution of a society's symbolic representations of the links among nature, society, and self. Where there is not adequate and shared symbolization of order in this sense, people are bereft of a vocabulary with which to articulate a legitimate account of what and how they wish to preserve, to control, to solve, and to change. The vocabulary of community becomes degraded into propaganda; the social (as against chronological) meanings of time itself become problematical; and there is confusion regarding by whom and on what bases it is decided how new tools are to be assimilated into the life of society. In this context the most significant form of cultural pessimism is that based on the expectation that the logic of mere physical survival–utility will provide the basis for new symbolic structures of order in place of earlier but failing paradigms.

The accumulation of social problems to the point of a potential species-survival crisis is generating a rhetorical conflict that will symbolically affect all existing ideologies and induce acute ambivalence in the moral reflections of all informed persons. That conflict is the contrast between two justifications of social order: the principle of human dignity and the principle of human survival. The first is embodied (for modern Western societies) in the vocabulary of abstract humanism and in the civil practices of liberal democratic societies. The principle of survival, on the other hand, is what primarily legitimates cybernetic vocabulary stressing the symbolic reconstitution of the world as a problem-solving system. The reconstitution of a world metaphor is a supremely religious, as well as social, phenomenon, affecting and affected by events in all secular social domains.

This chapter examines the conflict between the rhetoric of survival and the rhetoric of dignity as a conflict between two ways of justifying social order. It is not that survival and dignity are irreconcilable. They must be reconciled if we are to survive as both a biological species and something more than that. My intent in this chapter is to address critically the proponents of three intellectual positions:

1. Those who recognize no potential conflict between survival and dignity;
2. Those who recognize the conflict merely by recourse to the rhetoric of abstract humanism, on the apparent assumption that the meaning of human dignity is sufficiently self-evident that most people will act effectively in its name when challenged by social phenomena that portend human indignity;
3. Those who recognize the conflict on so pessimistic a level as to conclude that, in the name of dignity, all steps toward reconstituting the world as a problem-solving system must be opposed.

The moral ends of survival and dignity can indeed conflict, but not necessarily so. If they are not to conflict, certain steps must be taken. One of these is the conceptual clarification of what the ends of survival and dignity imply for models of social order.[1] Because the connections between survival and social order have been far more extensively explored in the general literature than have the connections between dignity and social order, this chapter will stress the latter. Indeed, my primary purpose is to salvage the concept of human dignity from the dangerously abstract humanistic rhetoric to which it has been entrusted.

The Problem

Are there any effective limits to what can be done to people in the name of their collective welfare? If there are not, then there are no barriers to the technological manipulation of persons so long as this is carried out in the name of social welfare. If there be such limits, however, whence comes their validity? And can one be sure that defending such limits is not to act against the public interest in the name of some romantic defense of parochial privilege?

To pose such questions is already to assert that there is more than one way of being human. There would be no problem if one thought without question that the "human" is essentially a biological fact and that sociology has as its sole object the study of the formal complexities displayed by large aggregates of highly intelligent organisms.

The most basic way of saying that being human is something more than membership in a socially organized population of intelligent organisms is to insist on the importance of the idea of a uniquely human dignity. If the dignity of the human means anything, it must refer to something so morally intrinsic to what one means by being human that its claims can be asserted against any momentary definition of social welfare. Talk about dignity—or dignity-talk as I shall call it—is speech about standards in light of which all models of social order can be judged. To deny the importance of dignity-talk, then, is to say that there is no moral meaning to being human except social forms and that any social order can be evaluated only in terms of its own survival or the efficient execution of its own stated goals. Dignity-talk presupposes that the human takes many forms, only one of which is the social. The social is something that we share with animals and insects. It is the collective form taken by biological creatures when we look upon them as organized populations. Humanity can manifest itself also in political, religious, artistic, and other forms, some of which we are driven to think of as more uniquely human despite difficulties we experience in defining exactly what we mean thereby. (We just know that chickens may have pecking orders but they don't have gods.) Humanity can manifest itself in highly organized form, leaving little room for personal demonstrations of human dignity, or the opposite. But in whatever forms we find humanity organized, systematic speech about social order always occurs in metaphors (organic, mechanical, theatrical, familial, etc.). Yet we sense that dignity-talk is about something that lies beyond the bounds of metaphors.

If dignity-talk, then, is speech about standards for evaluating the claims of social order, it is important to ask when and why such speech is or is not socially effective. Speech, after all, is more often than not socially ineffective. And its ineffectiveness is not necessarily rooted in the intellectual invalidity of one or another proposition. Validity is not necessarily power. Indeed, as I shall argue subsequently with respect to dignity-talk, the very need to speak polemically in the form of propositions may itself reflect a crisis of faith in the reality of that which is the object of discourse. It is this crisis that we must examine if we are interested in the pessimism that is so prominent a feature of many Western reactions to the advent of the modern technologically organized social order. It follows that the mere existence of dignity-talk as such does not in itself constitute a barrier to indignity. Nor does the intellectual validity of dignity-talk, however important this be, guarantee its social effectiveness.

If we are interested in dignity-talk as a mode of action (i.e., holding society to account and setting limits to the technological manipulation of persons), we must attend to the relevant aspects of this task. There are two in particular that inform my treatment of dignity-talk in this book: (1) the relation between the theoretical (intellectual) and the pretheoretical (experiential) dimensions of that which we speak; and (2) the distinction between those features of dignity-talk that are of purely intellectual interest and those that are of sociological significance. I wish to inquire into the degree to which modern culture (official categories of reality-talk) and society (structure and organization of social practices) can be thought of as "permitting" the exercise of dignity-talk as a contribution to effective moral action. Just as society can sentence people to death by physical extinction, it can sentence ideas to death by abstraction and irrelevance to social practice. Has this happened to the notion of dignity in any irrevocable way? How would we know? What kind of dignity-talk is of greatest significance for judging a technicist social order and for helping to set limits on its legitimation?

To proceed effectively along these lines it is desirable to begin with a conceptual analysis of contemporary dignity-talk. Only in this manner can one raise and pursue questions about the proper place of definitions, about the relations between intellectual validity and social significance, about the continuities between intellectual and experiential aspects of moral life, and about the place of metaphysical problems in the worlds of social practice.

The Concept of Human Dignity

Dignity, according to the dictionaries, means the quality of being worthy of merit, desert, honor, or respect. This, of course, is not very helpful. Let us thus begin by asking what it is we really want to know about the notion of dignity and its uses.

Dignity-talk is not rooted in abstract intellectual curiosity about definitions. It is inspired by one or another experience of indignity. Like justice, dignity is one of those terms that is not universally translatable yet has a universal resonance. There does not seem to be any society in which people are normally and collectively treated with the casual disdain reserved for the truly profane and inconsequential things of this world. Perhaps, then, an approach to the problem by way of a direct search for plausible definitions is not the best way to begin. For one thing, the traditional definition of dignity as "worthy of respect" has virtually no prescriptive implications except perhaps the general precept that one should respect that which is worthy. Furthermore, a definition of such generality does not indicate what precisely is the object of respect. More specifically, when we speak of human dignity, it is not clear what the referent of "human" really is. Is it the body? the mind? the ego? the soul? the personality? some special capacity? some collective social formation? More important still, the general definition of dignity does not tell us why something is worthy of respect. It tells us nothing, nor does it tell us how to find out what it is about the object that is worthy of respect. In other words, there is no total context in terms of which a claim that something is an object worthy of respect becomes intelligible to a person who challenges the claim by asking why.

Let us then examine three different ways in which the question of dignity can be addressed: the phenomenological, the constitutive, and the definitional approaches. This classification should be regarded as a simple heuristic device with which to organize a number of thoughts about a complex topic.

By "phenomenological" I mean those approaches that begin by defining events and acts that are recognizable as "indignities" and then proceed to analyze the properties of such experiences in order to isolate deeper layered implications of what is meant by dignity.[2]

Such an exercise leads naturally into the second, the "constitutive" (or "conceptual"), approach. Here the intention is to identify presuppositions and assumptions that make dignity-talk intelligible. Such efforts are conceptual since it is not necessary, in such an approach, to impute ontological status to these assumptions. The intelligibility of dignity-talk need presuppose only the acceptance (for whatever reason) of the constitutive rules from which can be logically derived the propositions of dignity-talk. Because of its importance for our subsequent discussion, this approach requires further attention here.

The concept of constitutive rules was developed by John Searle:

> Regulative rules regulate a pre-existing activity, an activity whose existence is logically independent of the rules. Constitutive rules constitute (and also regulate) an activity the existence of which is logically dependent on the rules (1969:344).

In Searle's usage, the notion of constitutive rule is logical. His most effective examples are drawn from games, and his subject matter is the philosophy of language. Thus, his concern is with the logical presuppositions of certain kinds of speech, not the sociological, cultural, or historical aspects of speech. Something like this method also appears in H. L. A. Hart's examination of the concept of natural rights (1955). Hart argues that rights-talk is intelligible only if there is acceptance of the one logically necessary presupposition to all such talk: the equal right of all men to be free. Charles Taylor (1971) has extended this line of reasoning to methodological questions in political science in a way more relevant to our own concerns. In Taylor's view these constitutive presuppositions are embodied in both a society's language and social practices, a distinction he regards as rather artificial. Thus, people need not be intellectually aware of the constitutive assumptions of their society. They accept them every time they act out socially instituted practices.

> The situation we have here is one in which the vocabulary of a given social dimension is grounded in the shape of social practice in this dimension; that is, the vocabulary wouldn't make sense, couldn't be applied sensibly, where this range of practices didn't prevail. And yet this range of practices couldn't exist without the prevalence of this or some related vocabulary. There is no simple one-way dependence here. We can speak of mutual dependence if we like, but

really what this points up is the artificiality of the distinction between social reality and the language of description of that social reality. The language is constitutive of the reality, is essential to its being the kind of reality it is. To separate the two and distinguish them . . . is forever to miss the point (1971:24).

Put in this way, this is a very strong claim indeed, and perhaps unacceptable on its face. But if one introduces a caveat, then the central importance of Taylor's point is preserved while the form of the claim is made more subtle. The caveat is that in many societies there is more than one language applied to social practices (the most immediately obvious example of this being the distinction between the language of participants and the language of scientific observers of a social practice). The full force of Taylor's argument becomes evident in his examples (the concepts of "voting" and "bargaining"). Later in the same essay, analyzing the concept of "legitimacy," Taylor calls these constitutive rules "intersubjective meanings." And he sharply distinguishes these from "subjective meanings" in the psychologist's sense of "attitudes." Intersubjective meanings, for Taylor, are not the sum of individual subjective attitudes. Rather, they are the assumptions implicit in the structures of social *practice*. In this sense, they are truly intersubjective, not multisubjective. One perhaps should add that to describe constitutive rules in Taylor's sense as presuppositions of practice does not mean they cannot be given intellectual form.[3] Indeed, Taylor does exactly this in his examples of voting and bargaining. It is, rather, to say that the primary *social* significance of such rules is not their manifest cognitive status in the minds of individuals but their latent status as presuppositions of practice. They take on intellectual form only when people consciously seek the deeper intelligibilities of their actions, which does not happen often.[4]

One further comment is required to adapt these considerations to our purposes before undertaking to define the meaning of dignity. The examples drawn from Searle, Hart, and even Taylor focus primarily on the logical relationships among constitutive rules, speech, and social practices. If, however, one is interested in social, cultural, and historical dynamics, more than logic becomes involved. The ontological status of constitutive rules becomes an issue, along with the sociocultural determinants of public opinion, motive, belief, and faith. What to the logician may be constitutive rules become, for the sociologist interested in the wellsprings of action, popular assumptions presupposing the intelligibility of existence itself. This must be kept in mind as we turn now to the problem of definition.

In approaching this problem it is well to recall that strong cases can be made by those who deny the dignity of humanity or who consider the whole question meaningless. But the stakes are too high for us to become discouraged. Dignity-talk, however indirectly, deals with the limits that can be set on the technological manipulation of humanity. In other words, dignity is a code word for whatever is considered to be intrinsically inviolable or "sacred" about the human status. Dignity-talk is talk about the limits of permissible profanation of the human. Hence, if dignity-talk is considered unintelligible to the modern intellect, then it must follow that there are no theoretical limits to the social legitimation of technicism. These are the stakes. Now to the problem.

Let us begin by eliminating from serious consideration two kinds of definition that do not, in my view, sufficiently reflect the direction of our concerns. This will enable us to move with better foundation to the requirements of a definition that does attend to these concerns. These two approaches may be called the "motivational" and the "instrumental." They are best assessed by way of illustration.

The motivational approach to dignity can be detected in the arguments of Ervin Laszlo. His most direct attempt to define dignity is this:

> Human dignity, I suggest, resides in the sum of the satisfactions of the human being: in the sum of the matchings of innate norms with the corresponding environmental states. Thus human dignity signifies the being (biological need–environmental matchings) as well as the well-being (cultural requirements–environmental matchings) of the person. An existence in which intrinsic requirements are matched with extrinsic conditions possesses excellence and is worthy of esteem (1971:178).

To understand what Laszlo is doing here, one must know what he means by "being" and "well-being."

> Both being and well-being can be viewed as states brought about by the matching of intrinsic norms or codes of the organism in the environment. These are normative organism-environment transactional states, representing specific input-output patterns. . . . My thesis is that both the biological being and the cultural well-being of the individual represent normative organism-environment transactional states, in which the individual's needs and demands are matched by the appropriate environmental states and events (ibid.:166, 167).

This transactional cybernetic view of dignity rests on the concept of need motivation: "The preservation of such transactional norms may well provide the basic motivation of behavior—cognitive as well as noncognitive—and define the basic 'need' of the individual (ibid.:170). Laszlo denies that this is "biologically minded behavioristic reductionism" on the ground that his notion of need includes cultural as well as biological needs.

My objection to this motivational approach is twofold: its particular kind of abstractness and its arbitrariness. All important definitions of dignity are abstract, but this cybernetic approach abstracts the uniquely human condition out of existence by subsuming it under general systems analysis. Despite all talk of culture, in the end the human is considered an organismic phenomenon and the concept of cultural "need" becomes a residual category for needs-talk that does not quite fit into an organismic view of man.

More important, there is an arbitrariness in Laszlo's use of the need concept. Most significant efforts in Western moral philosophy have sought to avoid the kind of relativism implied in the idea that dignity (or for that matter happiness, love and other such resonant notions) can be reduced to manipulated need satisfactions; to the contentment born of some homeostatic equilibrium between the human condition and its environments. It is not that men do not have needs. It is, rather, that dignity has usually been imputed to some aspect of the human-as-such, some unique characteristic beyond the arbitrariness of need and the possibility of satisfaction: something that cannot be reduced to motives but rather informs all motives.

Another approach congenial to the modern mind is the instrumental definition. Abraham Edel has apparently taken this tack:

> Dignity is now regarded as an ideal, like justice or well-being, and it is evaluated by the type of life it makes possible, the human purposes it helps achieve, and the human problems to whose solution it contributes. These purposes and problems constitute its psychological and social base; its function is the articulation of principles and the mobilization of multiple human energies and feelings, furnishing a direction for the achievement of purposes and the solution of problems both perennial and contemporary. . . . As a dynamic ideal it is united in a powerful way with the ideals of well-being, justice, liberty, equality, and in the efforts to remove the major discriminations and exploitations that have beset mankind. If the appeal to human dignity can play a large part in such enterprises it earns its keep quite easily (1969:238).

The major difficulty with this view, in my judgment, is that of the instrumentalist approach to experience generally: the danger of trivialization through the reduction of ends to means. If dignity is really nothing but a construct that serves as a means for the mobilization of effort to attain particular ideals, then—no matter how noble these ideals may be—dignity is reduced to their handmaiden. Furthermore, one's sense of human dignity becomes perhaps irrevocably tied to one's faith in the attainability of these ideals. Yet there are cases in history, both well known and unsung, in which persons have reacted with great dignity precisely when in deepest despair about the attainability of any other ideal. There are times when dignity, far from being a means toward any other end, becomes the only end there is. Why? Because there is an unarticulated connection for many people between their assertion of, or appeal to, dignity and their inchoate sense of truth concerning something about the human status itself. The instrumental approach seems not to allow for any crisis of dignity per se, apart from the particular ideals to which it may be instrumental. To see why this is so, we should perhaps reflect a moment on how the instrumentalist might respond to these objections.

The instrumentalist might reply that his approach and no other makes possible the avoidance of clashes between ultimate values. Suppose, for instance, liberty and equality are both considered sacred values. Is not an instrumentalist approach toward dignity helpful in distinguishing between such values, enabling us to stress one at one time, another at another time, depending upon what seems morally appropriate (i.e., congruent with our sense of dignity)? In this usage, dignity functions as a kind of indicator of our moral sense about the appropriate connections between our general moral traditions and the concrete situations in which we find ourselves. The trouble with this approach, again, is its triviality. When dignity-talk functions inchoately as a kind of psychic moral thermometer, it becomes little more than a cry of pain or a slogan for battle.

This connection among psychological reductionism, instrumentalism, and trivialization also is illustrated in the discussion of dignity by Jan Narveson, himself a utilitarian philosopher: "[Dignity] is not something opposed to utility, but simply the sense that one's utilities are to count— which is just what utilitarianism is all about. That is, dignity is not separate from one's other desires, but simply is the sense, or desire that one's other desires are to be counted" (1967:184). In Narveson's view, dignity reduces either to the "equal claim of all persons' desires to be satisfied" or to simply one more desire among others, worthy of no special weight in itself. The psychological reductionism inherent in such instrumentalism is revealed in Narveson's interpretation of his own argument:

> Indeed, it seems clear that our account of utility implies that a person's dignity, in the irreducible sense, should count for roughly as much as the person having it wants it to count for, relative to his other desires. That a person's dignity would be offended is *prima facie* a reason for not doing what would offend it; but which of his other desires is to count as outweighing his sense of dignity is up to him (ibid.:185).

What, in my view, is left entirely out of account here is the relation between theoretical and pretheoretical (i.e., taken for granted life-world) levels of discourse. Put another way, the connection is ignored between people's moral responses and what they assume to be fundamentally *true* about the world and humankind's place in it. In the instrumentalist approach, dignity-talk is a means and hence can be multifunctional. This precludes the possibility of a fundamental crisis of dignity itself; dignity-talk cannot reflect upon its own possible obsolescence.

Yet the challenge of indignity inherent in technicism requires this possibility of reflexive dignity-talk. Technicism entails the possible misuse of technological language in the service of what some philosophers and scientists regard as ontological claims about what humankind and the world actually are. Ontological arguments are truth claims. If dignity-talk is to be a mode of action, it must be prepared to engage in speech about what is true or not true about the world, however partially it is given us to know such things. Truth-talk about the world can sometimes legitimate indignity, not by changing our emotional sense of indignity but by persuading us that our emotional sense is irrational--out of tune with reality. This situation constitutes a fateful civil war of conscience. Before it a purely instrumental approach is helpless since those who prize philosophical integrity often regard themselves, in such a war, as invited to sacrifice precisely their utilitarian instincts in favor of the higher virtue of fidelity to truth. Such a conflict is a tragedy and should be avoided whenever possible. A purely instrumental approach does not contribute to this goal.

Some Requirements for a Definiton of Dignity

I should first acknowledge that dignity-talk as such is a Western phenomenon. Any definition whatever will be an observer definition, heavily influenced by the categories of Western discourse. There is no hope of scouring the languages of the world for terms that translate into dignity on some minimum level of universal equivalence. So perhaps we should ask why it is that one would even want to engage in dignity-talk on any level more universal than the avowedly culture-bound preoccupations of Western metaphysics.

The answer is best given by way of an analogous situation. Consider the concept of religion. Many, if not most, languages of the world do not have terms that translate into "religion." It appears that the historical emergence of religion as a specialized term was a response to a process of institutional differentiation and specialization that was associated with increased cognitive self-consciousness. Institutional specialization generates multiple mental worlds, multiple accounts of reality, multiple standards of legitimation, and so on. A specialized term for religion emerges when it begins to occur to people that religious categories are but some among a number of possible modes of discourse about the world. Specialized religion-talk, one could well say, emerges in an inverse relationship to the taken-for-grantedness of religious accounts of the world.

Yet, scholars and thinkers of all sorts have generally considered this insufficient reason for treating religion as a concept of less than universal relevance. Why? Because what we reflectively call religious categories arise out of types of questions and answers about experience that are of universal scope, in whatever form they may appear. We may say that these questions and answers revolve around a universal set of hermeneutic problems, two of which will immediately illustrate both the universality and the ultimately religious nature of such problems. These are the problems of *origins* (how did "we" come to be?) and *theodicy* (why is there suffering and evil?). Whatever the form of the queries or the answers, no known society would find these questions unintelligible. This is because they pose the problem of interpreting experiences that are of the essence of the human condition as such.

Can such an approach be validly applied to the concept of dignity, too? What experiences of such primordial quality generate the universal hermeneutic ground for our specialized vocabulary of dignity-talk? I believe that virtually all the answers that could be given to this question are ultimately subsumable under one concept: *moral significance.* Is there any record of a society in which the human status is regarded as essentially trivial? There are some in which the data of human experience are regarded as illusory; there is certainly no universal agreement on what the essentially "human" is or in what sense it is significant. But it is, I think, safe to say that the "destiny" of the human is everywhere considered (and institutionally treated) as a serious, not a trivial or profane, question. There is controversy, of course, as to why this is so. Answers can range from parochial formulations (e.g., stories of how the gods created "us") to universal observer interpretations (e.g., the social control functions of dignity myths). In an investigation such as this, the requirements for a definition of dignity will necessarily fall closer to the latter than the former sort. However, there are differences among sociological observer definitions, even those broadly labeled "functional." Some are more abstract, more crassly instrumental, more removed from the intuitive intelligibilities of universal experience, than others. It is to avoid such pitfalls that I shall pursue three more specific requirements of effective dignity-talk in relation to the significance of the human status as such.

1. *Universality.* Given the radical cultural variability in how the human status is defined, the first important requirement for effective dignity-talk is a definition that is reasonably universal without being so abstract as to be useless for concrete purposes. Let us note but one of myriad examples of how cultural parochialism can contaminate the search for such universality. It may have been noticed by the reader that I have tried to refrain from referring to "the human being." Instead, my vocabulary has stressed the notion of human status. My intention is to avoid conflating "person," "ego," "self," "subject," and "agent" into a single concept; a conflation that is central to the modern Western phenomenology of the human. This conflation has facilitated a persistent tendency in the West to define the human both too atomistically and too holistically. The definition is too atomistic because all these notions became, in recent centuries, associated with the organically observable individual. Yet it was also too holistic because, once imprisoned within the individual, these varied concepts were fused into a single, broad conception of the human as individual (atomistic egoism), and it became difficult to cite just what it was about the individual that was worthy of such dignity. This is one source of the cultic individualism (and social nominalism) of modern times.[5]

It is probably impossible, however, to achieve any universal definition of dignity that is not flawed to some extent by the cultural parochialism of the language (and its associated histories) in which the effort is being made.

Therefore, two related requirements of a good definition should be noted. Close attention needs to be paid to the distinction between a strong and a weak definition and to the problem of prescriptive content. These last two requirements merit further comment.

2. *Strong versus Weak Definitions.* A strong definition of dignity constitutes an almost automatic response to indignity. It provides an enclave of sanctity in a world of price, a prophetic staff in the court of power. A strong definition has almost automatic prescriptive significance; its roots are planted deep in the soil of institutions, its intelligibility is evident to intuition and intellect. Such a definition, however, is as little a product of intellectual fiat as of wishful thinking. Rather, it is in itself an institution. It is produced not primarily by philosophers but by the organic processes of history and society. Such a definition of dignity is an aspect of fundamental order itself; extreme intellectualization of the issue is but a sign of faltering confidence.

We must content ourselves here with the search for a weak definition. A weak definition of dignity cannot be self-evident, fructified by the soil of daily custom. It is more in the nature of a barrier against dehumanization; erected by strenuous mental effort and designed specifically to address intellectual (and rhetorical) tendencies that appear to legitimate, condone, or disguise indignity. All this is not to say that a weak definition of dignity is therefore an instrumental definition, fashioned to function merely as humanist propaganda. Rather, a weak definition is weak because it is minimal: it is a necessary but not a sufficient condition for an efficacious response to indignity. A weak definition of dignity is the fruit of a search for the minimal, intellectually secure foundation of dignity-talk. Such minimal intellectual requirements are not pretheoretically self-evident. They are not safely rooted in the soil of custom. If they were, such a self-conscious search for purely intellectual foundations would not be necessary. If the contrast is unclear, consider the Greek notion of the human as part of a comprehensive logic of natural development or the Christian view of the human as sanctified by the image of God. Contrast these orientations (and the sense of human dignity that attended them) with the modern perspective in which the truly "human world" is scientifically regarded as a domain of "mere symbols" secreted by intelligent animals—a world of language but not of nature, a social world but one unratified by divine law, a historical world but one with no intelligible telos. In a world so devoid of pretheoretical faith in suprahuman supports, a conception of human dignity cannot be a ringing declaration of dogma. It can be only a statement of possibility: the possibility that the darkest indignities of one's time may yet prove to be irrational and that dignity may not be incompatible with reason. For a weak definition to play such a historical role, however, it must be fashioned with our third requirement in mind: some relevance to the problems of political action.

3. *Prescriptive Significance.* The worlds in which "is" and "ought" were innocently fused have proven vulnerable to the acids of social complexity and critical intellect. Therefore, the requirement that a conception of dignity have some prescriptive significance does not mean one within which "ought" and "is" can dwell together without friction. What, then, does prescriptive significance imply in the modern context? It implies that the conception must not be so abstract that it can mean just anything in practice. To be sociologically relevant, a conception must allow for a reasonably operational standard for discriminating between practices that instantiate the conception and those that do not. With respect to the purposes of this study, three such politically strategic criteria are especially worth noting.

First, human dignity should not be defined *merely* as whatever conduces to the survival of man as a species. To do so would be to confound human dignity with whatever is considered functionally requisite for species survival. I stress this point because technicist reasoning and social practice are founded on species survival as the primary end. This whole study is based on the recognition that there can be tension between survival and dignity as moral ends and is dedicated to finding ways of using technologies without becoming technicists. Therefore, we need a perspective on human dignity that makes it possible to ascertain when dignity has been violated in the name of survival.

Second, dignity must be conceptualized in a way that represents an adequate challenge to the prescriptive efficacy that technicist ideology and practice can achieve in the organization of public life. To take a single but crucial example, consider the rhetoric of authenticity and liberation as an antitechnicist strategy. If this counterculture rhetoric becomes the conceptual cornerstone of dignity-talk, the road to a technicist resymbolization of man and society will be virtually unencumbered. Liberation *from* many social norms in the name of a cult of authentic selfhood is quite compatible with a technicist social order. If one is willing to define one's moral destiny in privatistic terms, leaving the task of running the public order to systems engineers, then a nontotalitarian politics of trade-off between private communities and state authority is conceivable. But if one's concern is with technicism as a threat to human dignity, the conceptualization of dignity must be such as to provide a basis for criticizing and redeeming social institutions beyond the antinomian criterion of simply getting them off one's back.

Third, rejecting a cult of the self is not the same thing as abandoning the notion that every self somehow partakes in the condition of being human. Therefore, a conception of human dignity addressed to the challenge of technicism should encourage a philosophical perspective on social science theory itself that attends constantly to links among social structure, history, and the creative agency of all selves.

The conceptualization of human dignity should be (1) susceptible to grounding in a diagnosis and critique of technicist trends in all social institutions; (2) capable of revealing the ways in which persons can and do become defined as morally superfluous to the public order (the primordial indignity); and (3) suited to helping redesign zones of personal and collective agency to create conditions more appropriate to the requirements of human dignity.

Perhaps we are ready now to consider more directly what such a conceptual approach to dignity might entail.

Toward a Definition of Dignity

It will be helpful initially to distinguish between some terms that are related in moral discourse and yet differ in ways that illuminate the special meaning of dignity that I wish to stress in this study. Two terms in particular come to mind that it would be fruitful to distinguish from dignity: honor and respect. For many purposes dignity, honorableness, and respect-worthiness are validly regarded as synonyms. How can they be usefully discriminated?

A distinction between honor and dignity has been drawn by Peter Berger, Brigitte Berger, and Hansfried Kellner (1973). Honor is the reward of efficaciously performed, status linked duties, whereas dignity, "as against honor, always relates to the intrinsic humanity divested of all socially imposed rules or norms" (ibid.:89). Honor is associated with a "hierarchical view of society [and] is a direct expression of status, a source of solidarity among social equals and a demarcation line against social inferiors" (ibid.:86). Honor and dignity, therefore, are different in an important sense.

> In a world of honor the individual *is* the social symbols emblazoned on his escutcheon. . . . In a world of dignity, in the modern sense, the social symbolism governing the interaction of men is a disguise. The escutcheons *hide* the true self. . . . In a world of honor, identity is firmly linked to the past through the reiterated performance of prototypical acts. In a world of dignity, history is the succession of mystifications from which the individual must free himself to attain "authenticity" (ibid.: 90–91; emphasis in original).

The implication here is that dignity-talk arises in a society in which the subjective self and the social order are no longer confidently intertwined. Dignity *per se* becomes an issue when the moral meaning of what it is to be essentially human is rendered ambiguous by social and cultural instabilities. Whether these instabilities are social (e.g., class wars or economic collapses) or cultural (e.g., conflicts between secular and sacred categories or radical pluralism of life-styles and values), they have in common the effect that people begin to suspect that their human interests may not be congruent with their social identities.

As regards the term respect, that which has dignity is, of course, in one sense by definition, respect-worthy. However, not all forms of respect connote an identical meaning of dignity. Man can impute respect-worthiness to himself in the same sense that he imputes it to any formidable species of animal. Yet the respect that he pays himself as king of the beasts is not the same as the respect implied in the phrase "human dignity." There seems something special about the intrinsic worth of humans that goes with the notion of the human as something that is less than divine yet more than animal. Why is this so? We respect an animal in a rather metaphorical sense, just as we "respect" the power of a hurricane. Unless we ascribe some sacred significance to an animal (in which case we do not really see it *as* an animal), what we really mean by respecting it is that we acknowledge its performance in the light of its limited potential. We "respect" a fleet-footed horse, a "majestic" lion, or a "graceful" bird. The respect-worthiness we impute to the human transcends this. The almost universally evident temptation to attribute divine qualities to extraordinary standards of human performance implies an unwillingness to circumscribe the conception of the human by purely "secular" (i.e., biological, physical, or other definitive) limitations. There is, traditionally, a subjective sense of awe connected with this very lack of definitive limitations to human potency. Human voluntarism is awesome, in other words, because it seems to reflect powers that overflow the finite properties of any specific human being defined as a determinate object. This view was given probably its first philosophically explicit formulation by Pico della Mirandola in 1487. Pico describes God as having given man "a share in the particular endowment of every other creature." That is, man is free of all lawful archetypes so he may trace for himself the lineaments of his own nature according to his free will. Thus, man is a creature "neither of heaven nor of earth, neither mortal nor immortal" (Mirandola, 1956:6–7).

The dignity of the human status, then, resides in the extraordinary human capacity for intentional creativity (and, of course, destruction). Humankind participates in that potency capable of *world* creation and destruction. Therein lies its mystery, its dignity, its awesomeness. This power, if it be assumed to exist, transcends utilitarian definitions of personal interests. Respect for it rests not just on how it is used, nor even on its recognition by any one individual person. Human dignity as respect-worthiness rests on the sheer factuality of human potency and on the assumption that to *be* human is somehow to share in this power for agency *regardless* of one's personal desires or merit. To no other creature of (profane) nature is imputed such potency for creative and destructive agency. The eclipse of honor is a crisis of one or another society; the possible eclipse of dignity would be an ontological crisis of the human status itself. Without dignity, man would be man as a purely biological object. He would not, in some important traditional and mythic sense, be human any longer.

These remarks should help us to comprehend the nature of the pessimism that afflicts many modern intellectuals regarding their civilization's capacity to preserve human dignity. To make this connection one must not think of pessimism merely as a personality trait, as a pattern of temperament, or even as socially oriented anxiety whose object is the fate of one or another society or institution. The appropriate orientation is best introduced by reflecting on a paradigm situation. The situation is that of a person who despite appropriate motivation, reasonable education, and freedom from political tyranny nonetheless comes to feel that what goes on in the world has no coherent relationship to his morally intended actions. (We must assume here a person whose intentions conform to the dominant values of his culture and who thus justifiably considers himself representative of the publicly professed traditions of his society.) If, for such a person, there is a sharp discontinuity between "self" and "world," among moral intentions, actions, and outcomes, then it is both the meaning of the world and that of his own agency in it that are at stake. Pessimism in this sense is a crisis of confidence in the efficacy of human intentionality in relation to the world in which one resides (or, indeed, perhaps in the very knowability and coherence of one's surroundings as a world).

Such pessimism questions the moral intelligibility of experience as a whole and, in principle, is ultimately incompatible with all rational claims to human dignity. This, I submit, is so for the same reason presented in our earlier discussion of human versus nonhuman respectworthiness: dignity claims make theoretical sense only when the privileged ontological status of humankind is pretheoretically taken for granted. By reflecting further on this connection between pessimism and the notion of ontological privilege, we may clarify our sense of what human dignity is and then be ready to proceed to more explicit definitional considerations.

Despite its vagueness, the notion of privileged ontological status is important for comprehending any discussion of human action within an ethical frame of discourse. What the term connotes is the assumed potency (not dominance) of human agency as a creative phenomenon in nature. "Will" (and its variations), as a technical philosophical concept is, of course, a product of Western metaphysical discourse (Bourke, 1964). However, there is a loose, minimal sense in which the concept has a more universal significance. It is unusual to find a society in which the purposiveness of a human agent, including the power both to help sustain harmonies and to engender chaos, is not accorded special respect. This respect is sometimes associated exclusively with Western technological hubris. But Western civilization is marked by extremes in this regard. Society, in a collective sense, is often regarded by Westerners as capable of achieving anything through the deliberate application of technology. On the other hand, Western culture also produces numerous theories about how helpless the self is in affecting the world (e.g., theories of inverse relationships between functional rationality and substantive rationality, theories of alienation, and theories of inauthenticity). In primitive societies with myth oriented cultures, one could almost say the reverse was true. In such societies there was far less emphasis on the omnipotence of collective technologies but a much greater sense of the self as agent. Indeed, this power was considered of cosmic and magical proportions. Hence, in such societies witchcraft was always an object of the most rigorous social control. Westerners sometimes forget that witchcraft beliefs reflect a profound confidence in the powers of individual human agency over natural forces. It is this respect for the power of human agency that is signified here by the phrase "privileged ontological status." (Naturally, the metaphysical explanations for this power vary greatly across societies, as do the prescribed ways of controlling it.) The truly terminal crisis of human dignity would be a total loss of pretheoretical confidence in the efficacy of both human cognition for comprehending a world and human agency for acting upon it. Such a loss of confidence *is* a loss of dignity because humankind would have to be redefined as primordially helpless, of interest only as one among the many crystalline shapes of matter indifferently created and destroyed by the morally blind process of nature.

Human dignity seems to require a philosophical anthropology appropriate to it: a "life-world" in which certain things are regarded as true on a pretheoretical level.[6] I previously said that the primordial experience that generates a felt need for dignity-talk is the sense of some threat to the moral significance of the human status and its place in nature. One must also ask, what kind of philosophical anthropology implicit in the culture of a people can support such a conviction of human significance. Any general answer must necessarily be very abstract since cultures vary so radically in content. But two stipulations can be made. First, there must be an implicit ontological assumption that the subjectively experienced self is "grounded" in some kind of "world" that (whatever the requisite preparations) is ultimately intelligible, accessible to human comprehension. Second, there must be an implicit ontological assumption that the self (whatever the requisite preparations) is potent in relation to its ground; that the self can act in or on the world in a manner informed by the world's intelligibility. In short, human dignity is based on the pretheoretical conviction that the self is grounded in a world from which it derives its significance; that the world is ultimately intelligible; and that the self partakes in potency to affect the world. All else is relative.

These general comments suggest the following definition of human dignity. Human dignity is the respect-worthiness imputed to humankind by reason of its privileged ontological status as creator, maintainer, and destroyer of worlds. Each self shares in this essential dignity (i.e., is recognizable as a moral entity) insofar as it partakes (whether by conscious intention or not) in world building or world destroying actions. Thus, human dignity does not rest on intention, moral merit, or subjective definitions of self-interests. It rests on the fact that we are, in this fundamental way that is beyond our intention, human. We are moral agents. Only suicide—the ultimate act of ontological self-rejection—can release us from that condition into which, to use Heidegger's term, we have been "thrown." The assertion of human dignity, then, is the constitutive act of moral consciousness. It is the entrance ticket to the community of formal moral discourse. To assert dignity is both to acknowledge the factuality of human creative agency and to accept responsibility for its use.

Given this definition, it now makes sense to ask under what conditions human dignity is likely to be taken for granted (possibly to the point that no special vocabulary for it exists). Four interrelated classes of conditions have been implied in the analysis thus far. These may be labeled phenomenological, psychological, theoretical, and sociological. Explicating them briefly provides an opportunity to summarize our reflections to this point.

PHENOMENOLOGICAL CONDITIONS

Human dignity is experientially possible in any culture with a pretheoretical structure of consciousness that supports it.[7] This means a life-world in which human agency is perceived as something that cannot be reduced exclusively to any deterministic account of humankind derived from observations of nonintentional phenomena. Different cultures have diverse ways of articulating such a basic phenomenology. In some, the human is considered partly divine. In others, the human is the highest (e.g., the most perfectible, or most complex, or most self-conscious) stage in the evolutionary dynamics of nature itself. In still other cultures, the human is that which is considered capable of self-determination and therefore can be spoken of "objectively" only in terms of metaphors (e.g., organism or machine).

Another feature of such a phenomenology is that human agency is perceived as of extraordinary, even unpredictable, potency. This potency is never trivial. It is, rather, on a world creating or world destroying scale. Accordingly, human potency, in many such cultures, seems to take on a numinous quality and become the source of myth and legend.[8]

PSYCHOLOGICAL CONDITIONS

The psychological correlate of the pretheoretical experience of dignity is a sense of awe in the presence of an intrinsically important, perhaps sacred, phenomenon. Thus, the respect imputed to human status is not merely an intellectual acknowledgment. It is a holistic experience involving awe and fear as well as other motives including, according to circumstance, identification, commitment, and obligation.

THEORETICAL CONDITIONS

Cultures vary widely in the degree to which pretheoretical assumptions, attitudes, and experience are reflexively articulated in intellectual forms such as philosophy, theology, or science. Theory is regarded by its producers as an outcome of collective, elite, transgenerational intellectual effort. These characteristics, it is understandably supposed, produce the most logically sophisticated truth claims possible in history to date. Since influential theories are generally reinterpreted until they are congenial (or irrelevant) to important social interests, certain types of theory eventually emerge as "institutionally sanctioned reality-talk."[9] When "theory" (in this general sense) becomes a major component of a culture, it takes on a kind of "official" status that lends certain ideas some independent historical influence in their own right. That is to say, theories begin to function as historical gatekeepers that help determine the intellectual respectability of pretheoretical assumptions and convictions.

In those cultures with a heavy theory component, therefore, human dignity requires both substantive theory and methods of theorizing that are compatible with it. If a society's dominant intellectual traditions downgrade the importance of human agency in their theories about the world or if dominant intellectual methods persistently ignore the requirements of showing connections between human agency and the structure of the world, then these intellectual traditions will contribute to the atrophy of human dignity. (Though they may, of course, contribute to the prevalence of dignity-talk!)

SOCIOLOGICAL CONDITIONS

Finally, there is the sociological component of human dignity. The social organization of a society can itself be such as to reveal or to conceal the facts of human agency (facts that include the generative consequences of personal action, the loci of opportunities for altering established structures, and the creative powers of language). Just as theory can "permit" or "impede" human dignity by structuring the way people believe and think about the world, so a society can be organized in such a way as to permit or impede people's recognition of their powers to act upon their world in morally consequential ways.

Resolution and Conclusion

I have thus far not spoken of the substance or scale of the worlds that are the presumed objects of human potency, just as I have not been specific about the notion of the human. This obviously is because of the variability in how the human and its worlds have been concretely conceived in diverse cultures and epochs. What, then, have we achieved with this more abstract approach?

Our procedure affords us the basis for an ideal-typical view of human dignity applicable to many varied contexts. Furthermore, it enables us to distinguish between strong and weak definitions of dignity according to the criterion of what is meant by the world in which the human presence is considered morally efficacious. Thus, for the ideal-typical strong definition of dignity, we must imagine a society (nowhere fully incarnated) in which the world that grounds the self is a seamless web incorporating all dimensions of human experience: physical, social, and temporal. In such a world, almost anything the self does or fails to do ramifies in a kind of sacred arc. That is, the pretheoretical interpretation of the world is such that all is interconnected. Everything—naming, technical activities, social interactions of all kinds—has potential ritual significance. Even the ubiquitous taboos surround the self with myriad reminders of human potency in the very structure of being.

As regards weaker definitions of dignity, here the worlds in which the human presence potently resides are much more restricted. In these worlds, experience is more or less fragmented and compartmentalized. The glorious claims of human potency have undergone major "humiliations." Our modern world has been racked, for example, by at least four such humiliations. These are the "Copernican," in which the earth as human world is cast out of the center of the cosmos; the 'Darwinian," in which the human is degraded to a stage in the evolution of the animal; the "sociological," in which the human story loses its mythical dimensions and is degraded to a function of social order; and the 'Freudian," in which human intentionality itself becomes a secret agent of the hidden beast in human form.

Can those worlds appropriate to a strong definition of human dignity disappear, rendering a strong definition historically obsolete? The answer to this is clearly yes. But a further question remains: is it possible for a world to emerge in which *any* definition of dignity, however weak, is irrelevant?

For those whose response to this question would be uncritically affirmative, human dignity is just another culturally relative concept like "witchcraft," "demons," or "phlogiston." What would be required to support a contrary claim: the claim that the concept of human dignity has timeless significance? It would be necessary to show that there is, in some irreducible and universal sense, a world that requires the active agency of all human selves to create, maintain, or destroy—a world that is a truly human product and in which all humans dwell. If such a world exists, then human dignity (in however weak a form), legitimately retains its place in the moral vocabulary of humanity.

There is such a world. It is *language* itself; a world for which all persons, intentionally or not, share some degree of responsibility because they all possess some degree of potency over it. This generalization does not, of course, apply to language alone. Humans produce a number of specifically human environments not reducible to language such as iconic, mathematical, musical, and gestural environments. However, for our purposes, language possesses a privileged position. Aside from gestures, the linguistic mode of symbolization is the only one in which all members of a society (except those who literally cannot speak) are actively involved. Although we always know more than we can tell, it is nevertheless through language that we all predominantly communicate. Therefore, it is the status of language as a form of human agency, and its relation to the possible meanings of "person" and "action," that presents the critical issue for human dignity in an age of symbolic pluralism, doubt, and disunity. As speakers we are often united, for through words we help create new forms of symbolic life that eventually dominate our collective conscience. As listeners we reside in a fully humanly produced world. And as interpreters of the speech of others, we become (however involuntarily) responsible moral participants in a community—the community of language. *How* we do these things is, of course, a desperately important question for those concerned with the empirical foundation for a moral philosophy of dignity. *That* we do these things is beyond reasonable doubt. We may believe that language is the voice of gods or men; we may admire the democratic flexibility or the authoritative traditions of language; we may claim our interpretative rights over language or delegate them (knowingly or not) to "experts." The fact remains that we each, through our partially but necessarily idiosyncratic understanding and use of language, help—in however small a way—to construct and alter the social world. And because of that fact, in this weak, minimal sense at least, we all possess the dignity of potency in relation to a world for which we are collectively and individually responsible.

It must be granted that the tension between survival and dignity as moral ends is a potential cultural conflict of truly great historic proportions, unlikely to be resolved in practice during this century. The issue involves questions of the highest theoretical magnitude that cannot be addressed within the existing academic division of labor. Nor can these questions be resolved by the sentimental pieties and bureaucratic tinkering to which contemporary liberalism has been reduced. The fate of language, our sole *common* instrumentality of world creation, is surely one of these great questions.

Notes

1. The philosopher who has most exhaustively attempted to develop a rationale for survival as a superordinate value on ethical as well as other grounds is

Pepper (1947, 1958). Edel (1960–1961) presents an exposition and critique of Pepper's position with which I substantially agree.

2.

> For dignity needs a way of expressing itself, of shining forth, of having a "sphere of free influence." Such an analysis of the concrete experience of "indignities" is merely an indirect approach to the full phenomenon of dignity. But before any more direct exploration it is of paramount importance to face the concrete situations out of which the present outcry for human dignity was born. This may not add up to a meaningful definition. But more important than a definition is the vivid experience of what it means to have the kind of intrinsic worth which is disregarded in dehumanizing treatment (Spiegelberg, 1971:61).

A similar methodological approach is used with respect to the concept of justice by Cahn (1949).

3. Searle comments on this point, too: "Can one follow a rule without knowing it? It bothers some people that I claim that there are rules of language which we *discover* even though, I claim, we have been following them all along" (1969:41–42; emphasis in original). For further explanation of Taylor's (1971) examples see note 4 below.

4. The possibility should be explored that the methodology appropriate to analyzing constitutive rules is perhaps phenomenology. Parallels between what Searle does, for instance, with the concept of "promising" (1969) and what Husserl meant by eidetic reduction (1962) although certainly not exact (e.g., one stresses conceptual, the other perceptual, reduction), are nonetheless intriguing and invite exploration. The same could be said of Taylor's method of explicating his examples. For instance, in his example of bargaining, he reduces the concept of bargaining, as a special form of negotiation, to constitutive assumptions that are really those of the market economy itself. (1971:23–24). These assumptions include "distinct *autonomous* parties in *willed* relationships," plus distinctions like "entering into and leaving negotiations" (emphasis added). Although Taylor does not carry out a conceptual reduction to the degree that Searle does with "promising," one could imagine from what he says how he might proceed. "Self-autonomy," "will," and "instrumentality" of social relations are but three of the constitutive rules assumed to underlie the practice of bargaining as contrasted with other forms of negotiation.

5. One of the best short discussions of the "human" as a problem in categorization is Mauss's essay on the person (1968). In an important commentary on this essay, Krader (1968) approaches the same problem from what is in one sense the opposite direction from Mauss. As he himself says:

> Mauss has advanced the notion of the person in its phenomenality as a category of the existent human spirit whose emergence is a cultural universal. His essay has been interpreted here as the opposite, not in its universality, but in its particularity, and in this sense as a phenomenology of the particular: the cultural-historic events of Western Civilization are the field of application of the concept of the person, which is then applied to intercultural phenomena (ibid.:490).

So, for Mauss the emergence of the person was a generalization of universal scope. For Krader the person is a Western concept that migrated from Western moral, legal, religious, and intellectual experience into Western technical science and other institutions of (now) worldwide significance. The issues raised in these pages constitute a profound hermeneutic problem for the whole notion of a social science.

6. The idea of the "life-world" is used here in the sense developed by Husserl, Schutz, Merleau-Ponty, and Berger and Luckmann. For a good, short treatment of the life-world notion that includes some of the problems that still require attention see Landgrebe (1940–1941).

7. The term "structures of consciousness" is not, of course, being used here as equivalent to subjective attitudes. Rather, it refers to a facet of culture: a structure of shared categories of meaning that function as cues, or grooves for thought, on the subjective level. The term is characteristically used by Nelson throughout his many publications on this topic. A good introduction to his use of a vocabulary for the study of structures of consciousness is Nelson (1964).

8. The term "numinous" is meant in Otto's sense (1967).

9. This is perhaps obvious with respect to social and psychological theories. However, it often appears to be the case also with physical theories. By way of example, some may recall the influence of Aristotelian physics on medieval Roman Catholic theology, of Newtonian physics on eighteenth-century social thought, and of Darwinian biology on nineteenth-century social thought.

PART II

Persons, Systems, and Environments: On the Sociology of Human Worlds

INTRODUCTION

One of the central features of technicist social thought is the importance of the concepts "system" and "environment." Most technicist talk of process is in terms of systems "adapting" to environments. Such talk is really about pseudoprocess insofar as it does not actually show how things happen or are made to happen by and to human agents. Those who would make sociology a science of *human* experience must examine environments as created for and experienced by humans.

Part I of this book closed with a philosophic exploration of human dignity rather than merely the conditions for systems or species survival. This account is rooted in the image of all persons as gatekeepers of language. As such, persons are participants in the making and unmaking of human worlds. This idea of world authorship through language is superficially easy to confuse with the notion that the world is a product of our "will." The concepts of "authorship" and "will," however, are not the same. Society is the sedimented actions of others (living and dead) that we experience as "traditions," "social structures," and "institutions." In these forms, society creates my "self" much more than "I" create society. The concept of authorship, therefore, must include consideration of the process whereby, through reflection, the "self" explores "within" for revelations of the presence of "others" in the structure of personality. Only subsequent to such reflection can the person be said to act rationally as an agent in light of the freedom that is open to him even as a social (and socialized) creature. One reason, perhaps, for the intellectual seductiveness of technicist thought styles is that they seem to provice a theoretical counterbalance to what has in recent centuries been an overemphasis in Western moral philosophy on behavior regarded as a direct product of human intention. Hence the attraction of the majestic impersonality of concepts like system, adaptation, and environment.

However, insofar as we cannot define in human terms what we mean by the systemic features of self and environment, these terms remain mere black boxes. As is the case with most (if not all) concepts, we are induced into the use of terms like system, environment, and adaptation via the connotations of specifically human experiences. "System" is implied by the world's resistance to our efforts to mold it according to our willed intentions. This resistance seems often to fall into regularly interlocking patterns that make us come to think of ourselves as opposed by a kind of complex object. Hence our eventual recourse to the idea of system. As everyone intuitively knows, these patterns of resistance come to be conceived as environments independent of our will with which we must come to terms (i.e., to "adapt"). The human experiences in which the concept of adaptation is grounded are surely the defeats, the humiliations, and the triumphs associated with compromise.

If the human sciences are to be spared enthrallment to linguistic technicism, we must labor to recover the experiential grounds of our abstract concepts. In Chapters 4 and 5 this is what I seek to do for the notions of system, environment, and adaptation. It is not that I would have us surrender these terms as useless. That would be to risk relapse into an overvoluntaristic tradition of discourse. What I hope to accomplish is to translate my earlier philosophic discourse about dignity into sociological discourse about authorship. To do this we should conceive of a human environment as a symbolic world into which humans are born as organisms and emerge, through socialization and reflexive consciousness, into the status of individual agents. Like organisms, agents rediscover ever and again the intractabilities of the world. But unlike organisms they intentionally fight back, they reinterpret, they lose, they compromise, and sometimes they win. This is so because the experiential nature of the world for human agents is something that can never be the case for organisms: the world for humans is a forensic context. That is to say, the human environment of persons is one of actions and events that often and persistently call for the reactions of condemnation and justification. Everyday life has some of the properties of a court of law in which persons are invited to think and act as counsels of prosecution and defense, judges and juries. One example of this, discussed in Chapter 4, is each person's position as an adjudicator of expertise claims on the part of those who would instruct him on the nature of things and what to do about them.

Human agents are authors of the social world insofar as they must interpret an ambiguous world through the use of language. To translate this general thesis into sociological terms we need to cope with at least the following two questions. First, what does it mean for sociological theory to speak of the world as a forensic context in general and of all persons as adjudicators in particular? My answer is to argue for moving the concepts of legitimation and delegitimation to the center stage of a nontechnicist formulation of a person centered sociology. If we are interested in the role of persons in world authorship, these concepts must be constructed in such a way as to make of legitimacy a highly dynamic process, not a social system constant. It should be noted, of course, that these remarks imply a usage of the term legitimacy different from what is usually intended by those who think of legitimacy as having primarily political reference. By legitimacy I mean the sense that persons seek to gain of the social world as being morally coherent (worthy, fitting, right, proper, etc.). Political legitimacy is only one aspect of such a quest. Seen in this way, legitimation and delegitimation are a kind of forensic master process engaged in by all normal persons in some way at some time in their lives. Indeed, in the microsense in which persons are constantly faced with breaks in their world's moral coherence, legitimation and delegitimation of actions and events are processes engaged in by persons each day of their lives. My elaboration of this approach and its significance for an understanding of the dignity claim of each person to be a participant in world construction forms the subject matter of Chapter 4.

In order to illustrate what it is I am criticizing in defining my task as I do, I begin Chapter 4 with a critique of two formulations of legitimacy that I reject. Both suffer from the sort of mechanistic, inadequately phenomenological reasoning that lends itself to eventual utilization in technicist models of social theory. In such models, the processes of authorship that comprise legitimation and delegitimation as mental and social actions are concealed. What this means I hope to spell out in my presentation and critique of the views on legitimacy of two thinkers, one a philosopher and one a sociologist. Subsequent to this critique, in the second part of Chapter 4, I turn directly to the task of asking what a nontechnicist sociological formulation of legitimation and delegitimation might be like. In this section I seek to give an adequately complex account of the various intellectual and methodological problems that must be addressed if a technicist formulation of the human world is to be avoided. In order to sustain the reader's sense of the relevance of these considerations to the aims of this book and to preserve the flow of the argument, frequent efforts are made to highlight such connections.

The second question that must be answered by those who would wish to translate a philosophy of dignity into a sociology of authorship is: what is it that persistently destabilizes the world such as to make of it a permanently forensic context? My answer takes the form of a conception of primordial scarcity, the subject matter of Chapter 5. Scarcity has meant different things to different thinkers. Part of the chapter reviews these diversities of meaning so as to allow me to locate my conception in a larger matrix of possibilities. The type of scarcity I have in mind generates empirical states such as structural contradictions, conflicts between moral and factual aspects of life, and discontinuities between modes of rationality. It is these states that in turn generate the activities that form the experiential ground of the concept of "adaptation."

In Chapter 6 I turn to a case study of what happens when these considerations of the human environment are obscured by careless recourse to black-box metaphors in the interest of controlling the "interface" between "systems" and "environments." The metatheoretical assumptions of physical science evolved out of notable successes in the study of physical nature and the prediction of its events, successes baptized by a phrase, the "conquest of nature," which would have stirred the most primordial anxieties in any other era. Flushed with victory, some ideologues of science began to treat all environments as potential domains for the imperium of their metaphysics. In some quarters this ambition has triggered primitivist counterrevolution against science; in others there has merely been resistance to the claim that a science of distinctly human activities is possible. If physical nature is "lost" to the cold majesty of mechanistic law, some would like to say, at least the human environment need not be "disenchanted" of its humanity. None of these extreme reactions is necessary if the dynamics of technicist mystification could be better understood. Chapter 6 is a case study of such dynamics. It examines the black-box uses of "cybernetics" in a variety of scholarly applications.

In essence, black boxes are conceptual shortcuts for getting on with the task of talking about some phenomenon when the nature of some part of that phenomenon is not adequately understood and when one does not want to get bogged down with that patch of ignorance for the time being. (Scientists were long willing to talk about electricity and its uses, for instance, without waiting to find out exactly what was the nature of electricity.) As a carefully controlled heuristic move, creating black boxes is a legitimate aspect of scientific inquiry. The trouble is that black-box creation seems to be a natural tendency of the human mind in the face of ignorance generally. It can become an almost magical resort to the creative power of language to conceive objects out of thin air. Technicism is based on this general human predisposition. Experts licensed (and expected) to produce "official" explanations are no less prone to illegitimate uses of black boxes than anyone else. Neither are scientists with careers that depend on publishing explanations arising out of contracted and limited research funds or politicians who traffic in stereotypes and simplifications for the sake of engineering votes.

What, then, is a specifically technicist mode of mystification from this point of view? It is the premature treatment of processes and actions as black boxes and the designation (and obscuring) of these black boxes through technological metaphors. What sort of technological symbolization most efficiently serves to conceal, rather than reveal, the processes whereby things happen? It would seem to be the phenomenology of switches and devices.[1] Concealment is a phenomenological property of switches as such, not a psychological property of the persons who use them. This concealment can never be lifted, for "if something is identified as a switch, the significance so revealed does not ultimately accrue to that thing, but passes through it to what the thing is a switch for" (Borgmann, 1972:138).

In a real technological system, it is possible to reconstruct the real process of which the switch is a part: to lay out the workings of the device that is shut on or off by the switch. In metaphorical usage, the device is either a conceptual black box or an abstract, often mechanical model that substitutes for a clear understanding of the real processes at issue. (Consider our everyday resort to questions and answers like "Why did Jane do that? That's her personality." "Why is there so much war? That's human nature." "What are those symptoms caused by? Oh, they're just psychosomatic." "Why did John break his leg three times in two years? He is accident-prone.") Even a real technological device has certain properties that make it different from ordinary things-available-to-persons-as-tools in several ways significant for a philosophy and sociology of technology. It will help in understanding Chapter 6 to lay out these sources of mystification in real technological devices.

First, a system consisting of a device and switch does not possess the immediacy of a thing like a hammer. Second, a device therefore has a large degree of independence from its context and from the talents and intentions of its users. Human labor and human ends are built (designed) into the very construction of the device. Once in operation, the device's virtue, so to speak, is to operate independently of future contexts except for the switch controller. Third, as an environment fills up with devices (i.e., relatively autonomous technological systems), persons begin to depend upon them. Hence, devices help create human needs that are aspects of the autonomous functions of devices rather than outgrowths of ongoing human aspirations. Finally, technological progress, up to now, has been largely defined in terms of the creation of efficient devices, an efficient device being a technological system that reduces demands on human energy, raw materials, and maintenance effort. To the extent that a population defines technological progress in this manner, however, it commits itself (largely unknowingly to be sure) to this process of reality concealment.

Such a phenomenology of concealment has causal implications for what might be called the institutionalization of personal incompetence. A tradition of over a hundred years of social criticism stressing the correlation between technological rationalization and personal incompetence stretches from William Morris to E. F.

Schumacher. In sociology this critical tradition is shared by thinkers like Max Weber, Karl Mannheim, Jürgen Habermas, and others who worried about the fate of "substantive" individual reason under the growing dominion of bureaucratized "functional" rationality. As everyone knows, the "natural laws" of classical political economy were regarded by Karl Marx as intellectual pseudodevices that conceal the praxis of human beings in history.

Chapter 6, then, is a study in the essence of technicism: the metaphoric use of technological imagery that, when not carefully controlled, fills the symbolic environment of persons not with real devices but with pseudodevices. Although the chapter deals specifically with cybernetics, in spirit it is about pseudodevices in the life of the mind. Chapter 6 thus is an inquiry into how the technicist imagination transforms groups and societies into "cybernetic" devices; how it turns the notion of personal and collective action into a system-environment "adaptation" problem; how it reduces the struggle for dignity to "system survival" functions; and how it regards the ambiguities of interpretive communication as a technological problem of symbol programming. Given this intention the reader will perhaps be puzzled by my reluctance, in the concluding section of the chapter, to give an unequivocally negative answer to the question: is a nontechnicist use of cybernetics possible? The apparent paradox arises from my hope that resistance to technicism need not be synonymous with fearful rejection of all technological metaphors in social theory construction. My cause is an attitude of self-conscious control of the metaphoric potential of language for the sake of disciplined illumination, not an anxious rejection of its demonic powers.

Note

1. In a complex and sophisticated analysis of the imagery of switches and devices, Borgmann has taken the first steps toward laying out a phenomenology of concealment through technology (1972). The subsequent discussion on switches and devices is indebted to Borgmann's effort.

4 The Structures of Doubt

THE FORENSIC CONSTRUCTION OF MORAL ORDER[1]

What does it mean for the world to appear to a person as morally intelligible or not? Of what general moral significance is it how this question is posed and answered on the level of technical sociological theory? Are some technical formulations of legitimacy incipiently technicist, and if so how? What have these considerations to do with the cultural and political reception of expertise claims in a social order? These are the constituent questions of this chapter on the human environment as created and experienced by human agents. In this and the following chapter I try to construct answers to these questions that can serve as criteria for detecting technicist standards of metaphoric borrowing from other disciplines, criteria to be applied to cybernetic systems analysis in Chapter 6.

Legitimacy: Implicit Technicism in Philosophy and Sociology

Let us begin with two conceptions of legitimacy, the critical examination of which will serve to introduce the problems that preoccupy us in Part II of this book. The two thinkers whose concepts in this area are to be looked into are philosopher Richard Taylor (1973:85–136) and sociologist Arthur Stinchcombe (1968:149–179).[2]

Taylor strongly distinguishes between legitimacy and justification: "The former are questions of law, and thus resolved by reference to rules, but the latter are questions of philosophy and resolved by reference to purposes or ends" (1973:87). This total separation between the factual (existence of

rules) and the moral (questions of justification) flows from three characteristics of Taylor's discussion that I want to argue against because I see in
them the ignoring of issues important for a nontechnicist understanding of
legitimacy. The first characteristic is what may be called a strong "brute-
fact bias." Second, there is too intellectualized a view of the nature of moral
justification in social life. Finally, Taylor's discussion takes insufficient account of the intimate dynamic relationship between persons and their political culture, an environment that exists both within and without a person.
To criticize these interrelated characteristics is to arrive at sorts of distinctions different from those Taylor makes. Specifically, where Taylor distinguishes only between legitimacy and justification and between government and individuals, I will argue for distinctions among legality, legitimacy, and justification and for distinctions among state, government, and
political culture. The few critical remarks to be made now will be elaborated in more affirmative terms when my own conception of legitimacy is
developed later in this chapter.

Let us turn first to brute-fact bias. For example, legitimacy on the level
of law, policy, or governmental order, says Taylor, "is a straightforward
question of fact and has no connection with philosophy, theology or
morals" (1973:89). The same is true with respect to political offices; "such
office, and the power expressed through it, are legitimately held provided
they are held in accordance with law" (ibid.:90). And, as regards government, "its necessary condition of legitimacy is that it exists, and that condition is sufficient" (ibid.:91). If a successful coup occurred, the new ruler's
power would become legitimate "at precisely the point when it became *in
fact* the supreme political power within that body politic. At that point, at
the point where this was the government *de facto*, it also became the source
of law itself, the government *de jure*" (ibid.:91; emphasis in original).

All this is consistent with Taylor's view that "in the realm of political
power there is no higher authority than the possession of power itself, for it
is only through this that anything becomes law in the first place" (1973:92).
And again, "The laws, policies and political offices of no *de facto* government can lack legitimacy, nor can any such government, considered as a
source of law, possibly be unlawful" (ibid.:92).

There is a sort of tough empiricist style to Taylor's analysis: "Government, in a word, is the coercion through threat and force of the many by the
few" (1973:92).

> Rules and laws do not govern, for they can compel or inhibit no one. . . . *Tyran-
> nical force, whether great or small, is absolutely inherent in the principle of self-
> government conceived as majority rule.* . . . There may be a justification for
> this, but if so, it has not yet been touched upon (ibid.:97, 108; emphasis in
> original).

These and other statements like them occur as parts of arguments seemingly
designed to disabuse readers of sentimental notions about the relevance of

moral-philosophical arguments to questions of political process and power. The irrelevancies Taylor examines include arguments from moral justifications of government, from utility, from popular sovereignty, and from social contract doctrines.

Realism in political matters is often a virtue, especially in the land of the tender-minded. But it can be carried too far, becoming unrealistically blind in its ideological toughness to the actual influences of moral reasoning upon political affairs. This may have happened here. Suffice it to say that social issues simply do not have that quality of brute facticity that Taylor's language suggests they do. Even laws, reflecting the effort to establish definitive, binding rules, are open to constant reinterpretation. The question of what the law is is seldom clear.

If social questions are constantly open to interpretation, then what Taylor calls justification is much more part of the warp and woof of daily life than his discussion allows for. There is a sense, of course, in which Taylor is quite correct. The problem lies in his restricted sense of the term justification. In his usage, justification is a highly intellectualized concept: the self-conscious intellectual construction, according to dictates of theory and logic, of reasons for doing or believing something to be right or wrong. Between legality (norm construction) and justification (intellectual rationale construction) there lies a realm of experience to which I shall later apply the term legitimation. In this realm persons struggle to clarify what they encounter as hazy patterns of moral intelligibility (or their absence) on the phenomenological level of direct experience. Only by recognizing in our metatheoretical assumptions the existence of this dimension of experience for all persons can we build into our scientific inquiries the sorts of investigations that reveal the active participation of persons in the construction of what, on more abstract levels, we experience as social facts.

These considerations lead us to Taylor's uses of the term government. Into this term he collapses what others might want to distinguish, such as state and political culture. It is his right. Why would one want to muddy the conceptual waters at the price of simplicity? The reason one might want to here is to take more effective account of the phenomenological realities of the legitimation process as experienced by persons as political agents. We see what a difference this makes when reflecting upon Taylor's critique of Socrates' account of why he will not escape with Crito and avoid his own death, an account Taylor uses as a major example of social contract justifications of state coercive power over citizens (1973:112–117).

In responding to Socrates, Taylor examines two different reasons for becoming a member of a society in order to establish the credibility for each reason of the contractual justification of state coercion. One reason is that of the fugitive who flees one society and accepts security in another in return for a promise to perform all duties lawfully imposed on him. This, says Taylor, is a validly contractual relationship. Taylor describes the other reason for being a member of a society in a fable of the wanderer who stumbles on a society that expects him to stay forever for very little guaranteed inducement (1973:95–96). This situation is much closer to imprisonment, "which cannot by the most farfetched imagination be represented as resting on contractual agreement." Efforts to make it do so are absurd. The fact that we are thrust into such a society from our mother's womb is irrelevant to the issue.

> Is our own situation [and Socrates'], then, like that of a party to a contract, who freely agrees to render to the other party certain things of value, out of consideration of the advantages to himself of doing so? Or does it more resemble that of a prisoner who has no choice whatever but to obey? Surely it is the latter, and the fact that we do not normally *feel* like prisoners shows only that we have learned to put up with things (ibid.:115–116).

To make his point as explicit as possible, Taylor ends his chapter on contract theories with an analogy to the status of a child in a tyrannical home.

> If one imagines a tyrannical father telling his ten-year-old that, in case he does not like the cruel and oppressive regime of the home, he is free to leave, and then interpreting the child's implied rejection of that "option" as his endorsement of the first alternative, the absurdity would be quite obvious. Interpreting a man's mere continued physical presence within his homeland as implicit contractual acceptance of its legal order is hardly less absurd (ibid.:117).

What might a modern Socrates answer? Perhaps something akin to this.[3]

> The house of my father is no more or less like a prison than is my self, in which I at times feel trapped. My father and the immediate form of his governance is at times (perhaps all times) cruel and tyrannical. But my father's house is more than my father and his relationship to me. His house, which is also mine, is more than a place, and more than his story or mine. It is the history of generations tied together by the idea of kinship, an idea partly mythical, partly factual. Whatever it be, this house that persists through time and the vicissitudes of history, dwells within me because it is partly what I am. It is no more (and no less) a prison than is life itself. My house is the source of pain and injustice. But as well is it the repository of the aspirations, the hopes and the rebellions that make of my house an idea, something yet to be achieved, as well as a prison-house of facts. As this is true of my house, thus is it true of my self. If my house be evil in my eyes, and I would be good, then I must leave it or become father in my turn and act in such a manner as to make of my deeds a living condemnation of all that preceded me. But either way, we speak not of a prison, for the soul is not like some disembodied ghost that but glides through a land of stones. The land is mine, and I am embodied in it. We must speak not of prisons but of a rambunctious kingdom, whose warring principalities are but versions of my self, and on whose throne I sit lone and insecure, yet responsible for this brief time for how my kingdom's history be fulfilled today and, through my actions, tomorrow until this kingdom be no more.

This imaginary response presupposes, on the level of politics, distinctions not made by Taylor; for example, state (the form of the polity, its constitution, partly myth and partly fact, that persists through time) versus government (the succession of temporal administrative practices).[4] Or consider also the implied distinction within my imaginary response between the self in its factual uniqueness and the self as equally factual participant in (and product of) the "collective conscience." For the former, society can indeed be a prison; in the latter, all crises of legitimacy are also little civil wars within the soul. Finally, there is a concept unexamined by Taylor that is necessary for comprehending my imaginary (and the real) Socrates: political culture. Political culture is neither purely fact nor purely idea. It is the arena of symbolic interpretations and acts pertaining to what "we" (the dead, the living, and the unborn) are all about as a collective community beyond kinship that persists through time as a historical presence in human affairs. Political culture is part of the constitutive order of the self; it is one form of how we experience the "right to say we."[5] We are born into a political culture. Our membership is voluntary, to be sure, but in the same primordial sense in which life itself is voluntary, "We," as members and agents of what we conceive to be the political community, do daily legitimate or delegitimate governments. We do so in terms of the telic ideals of the political culture as we experience them. And we do so by how we act or fail to act politically.

This process, as we shall see later, is not the same thing as conscious intellectual justification of reasons as the philosopher understands the latter. But neither is legitimacy to be dismissed as a simple encounter with the stone wall of brute fact. What the stakes are of this sort of metatheoretical discourse I will now attempt to clarify by taking up the same problem on the level of sociological theory.

Prior to examining his treatment of legitimacy, it is necessary to note Stinchcombe's (1968:149–179) definition of power. We will find here the same behavioral emphasis, the same brute-fact bias, the same lack of concern with *why* behavior is predictable—that is to say, the same separation between observed behavior and subjective reason—that we find in all social theorizing that lends itself to technicist utilization. Significantly, Stinchcombe designates his analysis as a "cybernetic" approach to power and legitimacy. By exploring it, we continue, in the context of this chapter, our line of inquiry into the uses of technological metaphors.

> By the power represented in some decision unit (an individual, a firm, a nation), we mean the *causal effectiveness* of the decisions of that unit. If my decision "yes" is a necessary and sufficient cause for the result to be "yes," and if my decision if not predetermined by my conditions of action, then I have complete power over the result. . . . We can say that a decision unit's power is the *amount of information added* about what is going to happen by knowing the decision of that unit (ibid.:163–164; emphasis in original).

We must ask what Stinchcombe actually means. It helps to reduce complexity by considering the simplest sort of situation: decisions to live or to die. Stinchcombe does this (ibid.:164–165), so let us examine his example.

He posits a man in a booth who can kill a man outside the box by means of an electrocution button inside the booth. But the man in the box knows nothing of the outside man's behavior. "The most power that the man in the booth can have is 'one bit,' the control over whether the man outside is dead or alive." If another button is added that flashes a light that says "sit down" or "stand up," and a light goes on inside when the outside man obeys, we may regard, says Stinchcombe, the man inside the box as having "two bits" of power. The point is that "the power of the man in the box, given his control over the motivations of the man outside, is directly proportional to the information-carrying capacity of the channels between him and the man outside" (1968:165). What is interesting here is the assumption of motivation control through life and death decision control. The only escape clause allowed from this assumption is pathology. "If, however, the man outside has suicidal impulses, the inside man's control decreases because the signals are no longer effective causes" (ibid.). There is no inquiry into the subjective reasons for behavioral predictability; what is apparently taken for granted, instead, is a sort of law of self-preservation.

What are we commiting ourselves to if we accept this reasoning? Let us see by varying the example. Posit a man holding a gun on another man and giving him an order to be obeyed on threat of immediate death. Presumably this man has near total power over his victim since they are in full view of each other and power "is directly proportional to the information-carrying capacity of the channels" between these two men. Here, it would seem, is a simple binary decision for the victim: to be or not to be. There is a high probability of obedience. Yet, those knowledgeable about the world, if they ignore their immediate intuitive fear of death, can perhaps imagine (or recall) situations in which there might be a high probability of *disobedience*. How are we to understand such counterintuitive situations? By understanding the wide range of meanings that the notion of self can have to human beings. Only thus can we comprehend what sorts of definitions of self-survival can supersede the individual-organic notion of the self that is preserved when a person obeys the holder of a gun.

How does this square with Stinchcombe's information conception of power? If the gun wielder were telepathic, then presumably the meaning of information would be coterminous with everything that could possibly be known about the victim's wellsprings of action. Short of this, we may be in the presence of a tautology (whatever I get you to do at my command reflects the efficacy of the information I have about you and the "information-carrying capacity of the channels" between you and me). For these and other reasons, we should critically question the information concept of power. To see why, let us turn directly to Stinchcombe's notion of information.

"For our purposes," says Stinchcombe, "the most convenient way to conceive of *relevant information* is the *degree of covariation* between the signal received by the power-holder and the variable he wants to control" (1968:171; emphasis in original). What are the factors that must be controlled in order to clarify this relation of covariation? It is necessary, according to Stinchcombe, to avoid:

1. Irrelevant information coded and transmitted from the source.
2. Bias in the encoding process.
3. Bias or adding irrelevant information in the channel from source to receiver.
4. Failure of the channel to filter out noise and irrelevant information.
5. Failure of the receiver to decode the information correctly, so that he misinterprets the meaning of the communication received (ibid:171–172).

The question is: what are we to understand by "irrelevant," "bias," and "misinterpret the meaning of?" These terms suggest, by implying their opposites, that there is such a phenomenon as absolutely relevant and unbiased information, which, if properly transmitted, can be interpreted in terms of some singular standard of precise accuracy. If we ask, relevant and unbiased with respect to what? the answer implicit in Stinchcombe's whole analysis is predictability and control of response. This is a technological criterion of meaning. It should not be surprising that Stinchcombe argues that "administration is above all information processing" (1968:164). It is a short step to the view that "commercial activity, quoting prices for thousands of goods and services, is thus essentially administrative activity for the economy as a whole. By considering this case, we can see some of the crucial features of administrative systems in relation to power structures" (ibid.:166). I shall not pursue this facet of Stinchcombe's analysis but turn instead directly to a consideration of what all this implies for him regarding the meaning of legitimacy.

For Stinchcombe, "A power is legitimate to the degree that, by virtue of the doctrines and norms by which it is justified, the power-holder can call upon sufficient other centers of power, as reserves in case of need, to make his power effective" (1968:162). This is presented as an "attempt to define the concept" of legitimacy. It does not, of course, tell us what legitimacy is in any subjective or objective normative sense: as a concept, legitimacy, for Stinchcombe, "tries to locate the causal significance of the contingent probabilities of actions of other power centers" (ibid.:163). Why look at it this way? Because "the chief significance of the legitimacy of powers for the social system as a whole is that it makes the powers available to a decision-making apparatus predictable, in spite of any opposition the exercise of power might run into" (ibid.:162).

What, then, of the phrase "by virtue of the doctrines and norms by which [power] is justified" in Stinchcombe's definition of legitimacy? It is the legitimacy of a claim "in the centers of power which is crucial, and not its legitimacy among people who must take the consequences" (1968:159). Accordingly, "A legitimate right or authority is backed by a nesting of reserve sources of power set up in such a fashion that the power can always overcome opposition" (ibid.:160). Thus for Stinchcombe, "The variable we want to define [as legitimacy] does *not* involve the *action* of third parties as a causal variable, but rather the *probability* of action of third parties in given contingencies" (ibid.:163; emphasis in original). This probabilistic, behavioral approach to legitimacy is explicitly designed to minimize references to subjective factors in favor of social structural factors. It is not surprising, therefore, that "justification" is seen by Stinchcombe as the effective application of abstract "doctrines of legitimacy."

> The crucial function of doctrines of legitimacy and norms derived from them is to create a readiness in other centers of power to back up the actions of a person with a certain right. Doctrines of legitimacy serve the crucial function of setting up that nesting of powers which usually makes appeals to physical force unnecessary (ibid.:160–161).

The concept of subjective factors is reduced to notions like public opinion, popularity, and willingness to obey, which are seen as "reserve sources of power" (ibid.:161).

Why might we want to object to all this? After all, is not Stinchcombe reacting against the psychological reductionism rampant in American social science? Does he not wish to assert the primacy of social structural analysis? The problem is not with his intentions but with the facile acceptance (in order to dismiss it) of the meaning of subjectivity in the crudest psychological sense. When subjectivity is thus vulgarized and dismissed, social structural concepts become reified, even as the complex dynamics of personal agency in the construction of the social world are left out of account. This combination of tendencies makes it seem almost pointless to inquire into the rationality of public opinion and of legitimacy claims. That sense of pointlessness is an important condition for the institutionalization of a mass mentality of technicism.

It should be added, too, that this approach misrepresents the efforts of those who wish to introduce a less crude conception of subjectivity into sociology. For example, Stinchcombe says of Max Weber that "he defines the legitimation of power . . . in terms of the acceptance by subordinates of the rights of superiors to control them." He then comments that

> in his concrete analyses of power phenomena, Weber was very little concerned with any estimations of the state of public opinion or of the ideological enthusiasm of subordinates and subjects. Rather, he analyzed the reactions of other centers of power. I think Weber's instinct in the analysis of concrete cases was right, and that his own theoretical analysis of charisma and its differences from other forms of authority should have led him to see that his definition of legitimacy would not work properly (ibid.:161, n.6).

The problem here is that Weber's project of introducing considerations about subjectivity into sociology had to do with far more than simply momentary states of public opinion or "ideological enthusiasm." It is true that, as Alfred Schutz (1967) took great pains to point out, Weber failed to develop an adequate systematic phenomenology of meaning to undergird his work. But even so, Weber's treatment of subjectivity is far richer than Stinchcombe acknowledges. There are at least six different levels or dimensions of subjectivity operating in Weber's empirical work. Only one of these is doctrines of justification (explicit rationales in Taylor's or Stinchcombe's senses). The other five we may briefly describe as follows:

1. The *immediate sense of* (immediate, unreflective, cognitive responses to specific stimuli)
2. *Opinion* (uncultivated, unreflective attitudes in the form of proto-doctrines)
3. *Ideology* (rationales of justification whose essence is that they are the naive reflection, on the level of ideas, of the way the world appears from a particular social location)
4. *Social logics* (the conceptual assumptions that underlie concrete social practices and that are intersubjectively acted out by all who engage in these practices whether they are aware of these assumptions or not)
5. *Cultural logics* (explicit and latent implications of cultural symbols that are implicitly available as sources for new interpretations of reality, motive, or action)

Any analysis of legitimacy based on a concern for the active agency of persons must be grounded in some such sort of sophisticated phenomenology of subjectivity. This general approach will be developed further in the third part of this chapter. I wish to end this first section by noting that neither Taylor nor Stinchcombe provides us with any conception of legitimacy that helps us use sociological analysis in the service of a moral philosophy of persons. In their hands, the concept of legitimacy inspires little more than inquiry into how social arrangements, abstract doctrines, and general attitudes can be technologically utilized to get people observably to obey other people.

Toward a Nontechnicist Sociology of Legitimacy

If the world surrounding the person is experienced by him (and conceived by the sociologist) as essentially brute fact, then it is a world relatively impermeable to personal agency of the sort required by a moral philosophy of dignity. In this part of the present chapter my emphasis is on the problem of legitimacy as it must be conceived if every normal person is to be regarded as a moral agent. I shall not be constructing a sociological theory of legitimacy. Rather, my concern will be with indicating what such a theory should be about. In Chapter 5 the focus shifts more directly toward those properties of the human environment that make it a generator of legitimacy problems.

Legitimacy can mean many things. In its most restricted sense, legitimacy refers exclusively to rules of legal and political procedure. In its most general sense, legitimacy can refer to standards of truth itself.[6] I shall deal here with legitimacy as the concept that summarizes those speculations addressed to the following general question: what is it that makes organized society possible other than the resort to physical force? Charles Taylor has usefully reminded us that no matter how political legitimacy is conceived

> it can only be attributed to a polity in the light of a number of surrounding conceptions—e.g., that it provides men freedom, that it emanates from their will, that it secures them order, the rule of law, or that it is founded on tradition, or commands obedience by its superior qualities. These conceptions are all such that they rely on definitions of what is significant for men in general or in some particular society or circumstances, definitions of paradigmatic meaning which cannot be identifiable as brute data (1971:35).

Thus, "we all know" (without normally being quite sure why) that it is possible for any given law to be experienced as morally improper even though successfully promulgated in a purely legal sense. Typically, codified norms are subordinate to the influence of the larger social order of custom.[7]

For many years in American sociology the textbook account of legitimation was based on the concept of socialization, i.e., the introjection of society into the inner domains of the psyche via learned norms, roles, and values. In recent years a debate emerged over what has been called "the oversocialized conception of man" (Wrong, 1961). As so often in intellectual history, it is partly the practical events of life that have compelled this reopening of theoretical questions about the meaning and sources of legitimacy. Let us briefly note three such connections.

The widespread crisis of "authority" in the Western world is now almost a cliché. Symptoms of this alleged crisis include the rhetoric of generational conflict and the present ideological ambiguity of behavior that was once thought of simply as "deviant" or even "criminal" (e.g., divorce, drug taking, or homosexuality).

A second example of the influence of practical events upon theories of legitimacy is the unpredicted ways in which people have responded to circumstances that contradicted norms and values traditionally thought of as safely "internalized." The most glaring example of this, of course, is the capitulation of masses of people to the demands of totalitarian regimes.

Third, under conditions of modern cultural pluralism the concept of legitimacy, much like the notion of deviance, has itself come under ideological attack (e.g., Wolff, 1970). Some argue that the concept of legitimacy is not scientific but inherently ideological, a tool of agents of the status quo.

At its extreme, this argument holds that tolerating a wide range of behavior heretofore under excommunication as "illegitimate" expressions of human instincts will help bring about a utopian condition of liberation and happiness.[8] Theories of legitimacy thus continue to have unavoidable ideological implications as they become embroiled in the old and new social conflicts of our time.

With this brief review of the practical stakes in mind, let us examine the concepts of legitimacy, legitimation, and delegitimation in social theory. Legitimacy is regarded here as a sense of the fitness (i.e., rightness and propriety) of one's human world. By this I mean the institutions, rules, and procedures in terms of which one discovers oneself to be related to society. I will later be concerned with how one may analyze where this sense of legitimacy comes from, and how it can be undermined. Legitimation and delegitimation are here regarded as dynamic interpretive processes—a struggle for intelligibility wherein persons seek an acceptable moral interpretation of the human world. I shall try to provide a partial account of why the human world cannot be definitively experienced as legitimate and hence why the struggle to legitimate is a permanent feature of what it means to be a person.

THE CONCEPT OF LEGITIMACY

As scientific observers, how may we speak of persons experiencing something as legitimate when they are in the mode of reflection that some sociologists call the "natural attitude" of common sense?[9] The scientific analysis of commonsense perspectives and behavior is a fairly new agenda in sociology. In the study of legitimation phenomena, the history of the present century amply illustrates the need for a more sophisticated sociology of commonsense experience.[10]

For persons to perceive something as legitimate or illegitimate is often a profoundly disturbing experience. In many situations, to perceive something as legitimate or illegitimate calls forth a sense of obligation to act against one's immediate interests, to refrain from acting for one's interests, or in some other conflictful way to depart from one's routine activities. Thus, social scientists must face a number of questions. Do people's acquiescence to a government's policies indicate support, indifference, ignorance, or fear? Does resistance to someone's actions reflect the imputation of illegitimacy to the acts or to the perpetrator, or does it mean something else entirely? There are those who argue that such questions about the phenomenological contours of action are unimportant because we need know only what men "do" about their perceptions, not what they "feel" or "think."[11] But, apart from the overrigid distinction between "objective" and "subjective" dimensions of action implied by such an argument, this view neglects the temporal nature of action. A person's failure to engage in observable action at any given moment can mean many things. Some of the things it can mean include what to the person himself may appear as "illegitimate" motives (cowardice, ignorance, insensitivity). Such self-judgment can lead to inner dialogues having manifest results only much later in time (results such as sudden outbreaks of revolt, acts of reckless risk, dramas of unexpected self-sacrifice, suicide, or less dramatic events such as slow changes in aggregate public opinion).

One of the benefits of phenomenological analysis in sociology is that we are gradually being forced to cease dividing everything that persons do into the two categories of "behavior" (observable motion) and "attitudes" (subjective orientations imputed on the basis largely of verbal behavior). Legitimacy is not just a subjective belief, opinion, or attitude "brought to" an otherwise objective situation by a person. Legitimacy, rather, is a facet of the whole experience of an object itself.[12] Legitimacy is that about an object, phenomenon, or situation which makes it appear morally appropriate (in one or more of a variety of senses such as "worthy," "right," "good," or "proper"). A complete phenomenological analysis of legitimacy as experience would be difficult and complex and will not be pursued here. But two important generalizations may be ventured.

First, the widespread intellectual emphasis on the distinction between subjective attitudes and objective behavior has obscured the fact that, on the simplest level of commonsense awareness, people experience many objects and events in an immediate sense as either true, right, proper, reasonable, sensible or not. For example, it obviously is sometimes a question of personal disagreement whether a specific act of killing is legitimately definable as "self-defense" or as "murder," and there is no harm in labeling some such quarrels as subjective differences of opinion. In all societies there are persons (e.g., elders, judges, legislators) who are empowered to have some latitude in defining the legitimate boundaries of terms like murder and justified homicide. However, imagine a Western legislature suddenly defining the months of July and August as open season for the hunting of humans for sport as what are now called "game" are hunted. It is safe to say that a Western public would not be likely to accept this policy as legitimate just because it was legally enacted or as simply a matter for disagreement on the level of subjective opinion.[13] Likewise, to take a more prosaic example, we do not—in the presence of a beautiful woman—tend to say, "There stands a human female," to which remark we then deliberately append statements like "I, personally, out of my subjective arsenal of assessment categories, add on my opinion that she is attractive, desirable, and elegant." We do not respond in such a way because the structure of our percept is unified to begin with ("*There* is a beautiful woman!"). On this initial level of perception, aesthetic and geometrical and other categories are often fused.

What do these simple examples tell us? They suggest the following generalization: in every society persons perceive some objects, events, situations, and so forth as *intrinsically* right or wrong, coherent or meaningless, necessary or optional. Legitimacy is an experience referable to an object, not simply an opinion derived from an arsenal of assessment attitudes. Of course, to uncover *what* is so experienced in a given society, by what sorts of people, and under what sorts of conditions should not be a matter simply for speculation. The answers to such questions should be determined through empirical research.[14]

Let us now turn to our second generalization about legitimacy as experienced in the commonsense mode of awareness. I have said legitimacy is that about an object, event, or situation which makes it appear morally intelligible. This formulation obviously begs some questions. In particular, three such questions should be noted: (1) what stands out as a context for moral judgment, in what circumstances, and to whom? (2) under what circumstances in daily life does it seem to become important to make a consciously articulated issue of the moral appropriateness of something? and (3) what aspect(s) in particular is it about an object, event, or situation that imbues the whole with moral intelligibility?

Consider a situation in which all three questions clearly arise. As is well known, from the legal standpoint a good deal of minor tax cheating occurs among Americans without, in most cases, such acts being perceived by their perpetrators in a moral context. Let us posit a married couple conversing on how best to evade reporting certain types of income. Now imagine their twelve-year-old son entering the room, indignant at what he has overheard, ready to challenge the moral credibility of his parents. In the ensuing conversation it is not difficult to imagine the tax evasion issue as being transferred quickly from a technical context to a moral one. The fact that it is their preadult and inexperienced son (rather than a close and worldly friend) who has thrown down the gauntlet partly determines the level of conscious articulation to which this particular moral issue will be driven in the conversation. Although to a friend one can say "come off it," it is easy to imagine a conflict of formalized legitimacy criteria emerging in an encounter between hapless parents and the Savonarolas that outraged innocence can produce. The son will probably insist on the criterion of impeccable lawfulness for morally evaluating his parents' discourse. The parents may perhaps defend their search for tax loopholes with a justification based on the government's alleged moral turpitude in misusing hard-earned tax money for wasteful bureaucratic activities, wars, and welfare programs. Such examples ought not to be dismissed as variations in "mere rhetoric." People do not encounter the same issue in the same way in all situations. Rather, we frequently experience different contexts as different sorts of invitations to reflect, more or less deeply, on the moral implications of the positions we take on given issues. We can be, and often are, of quite different minds on the same issue depending upon what is expected of us or whom we are addressing.

Space precludes exploring these reflections in the detail they warrant. However, a few more comments pertaining to the third question are in order. That question, it will be recalled, was: what aspect(s) in particular is it about an object, event, or situation that imbues the whole with moral intelligibility? The simplest level of constituting something as legitimate or illegitimate, as we saw, is its morally unreflective acceptance or rejection. There are circumstances that lead us to question such automatic perceptions. Whatever may be the everyday expression of such questioning, its basic phenomenological form is expressible in one or another of these three questions. What in particular is it about X that imbues it with legitimacy for me? What is absent here that leads me to be so uneasy about the legitimacy of X? What criteria of legitimacy seem to be in conflict here such that I feel ambivalent about X? What I am asserting, then, is this: only on the most unreflective level of commonsense awareness is it the case that a whole object, event, or situation is perceived as legitimate or illegitimate in an undifferentiated way. To move beyond this simple level of cognition is to recognize that legitimate objects, events, or situations are not integral moral unities by nature. They have certain features that appear to impart legitimacy to them as perceived wholes. (These features have variously been called values, legality, doctrines, charisma, etc.) Furthermore, all societies apparently have instituted criteria for distinguishing among more and less valid standards for the determination of what is legitimate and illegitimate. Cultivation of these standards of judgment is in the charge of those persons designated (by dint of birth, inspiration, study, magic, achievement, or whatever) as competent in their use. We may generically refer to such universal guardians of competence as "the experts." The moment that people embark upon moral reasoning, they are in potential conflict with such official guardians of cognitive virtue. Since it is in this way that the problem of technicism is experienced on the mass popular level, we should pause to look at the notion of expertise directly.

EXPERTISE: THE UNIVERSAL DYNAMIC OF COGNITIVE AUTHORITY AND REVOLT

It has been argued here that the practices of organized social formations (governments, churches, nations, bureaucracies, etc.) seem somehow to derive their validity or propriety or coherence from the laws, constitutions, doctrines, rituals, symbols, and myths associated with these formations. The word "association" hides, as we saw, a host of questions because the word does not signify a merely casual correlation. Rather, by association I mean a kind of symbolic irradiation whereby social organizations and practices derive their aura of legitimacy from certain symbolic sources not intrinsic to the formal properties of these organizations themselves. Whatever else irradiation may be taken to mean, it is at minimum a process of cognitive interpretation, no matter how automatically or unconsciously executed. The acceptance of "expertise" may be defined as the conscious or unconscious delegation of one's cognitive authority over some part of the world to persons one regards as more competent than oneself in the exercise of cognitive judgments. It comes as no news that conflicts now exist in virtually all modern industrial democracies between persons making knowledge claims on the authority of common sense and those making claims on the basis of credential expertise. What is not so obvious is the universal theoretical basis of the hostility expressed in modern civilization as the revolt against technocracy.

How are expertise claims normally legitimated in societies? The answer is that certain knowledge claims are asserted about how things work, and people are encouraged to accept as experts all those who are traditionally regarded as having proficiency in such knowledge and its uses. As was noted by both Adam Smith and Karl Marx, people seem to have a natural tendency to reify the division of labor. That is, people tend to identify the qualities, moral worth, and authority of a person with that person's specialized function in society. If a society is highly differentiated in its division of labor, more and more of life comes under the domain of putative experts. Thus, just as I am ill because I have what the physician defines as my disease or am nervous because I suffer from what some psychiatrist defines as my neurosis, so my national interests grow to be what my government announces them to be, and the level of my education is what my diploma says it is. To the degree that this assumption about the exclusive competence of ever increasing numbers of experts to define reality is accepted by a public without question, to that degree may we speak of a technicist culture.

Now, most traditional societies were mildly technicist in the sense that criteria for expertise were heavily institutionalized. But, in many of them, so were the constraints against misuse of expert authority. (Think of the heavy penalties in most primitive societies for using witchcraft powers for personal aggrandizement or the almost universal protection of the religious status of man in nature.) When human dignity is interpreted as the inviolable interests of the individual person, while at the same time the religious beliefs that undergird man's special status in nature are replaced by secular science, technicism becomes the transcendental moral problem of civilization. This is because the demands of moral individualism coexist rather uneasily with the deterministic, amoral, and impersonal picture of the world presented by modern secular science. Under such circumstances, a more self-conscious and critical orientation toward the legitimation of expertise becomes morally essential. Presumably, rational people in a democratic civilization, even in the name of antitechnocratic sentiments, would not want to do away with the concept of technical expertise itself. What is required, then, is for everyone to become conscious of this question: under what precise conditions is the delegation of cognitive authority to experts an essentially technicist act? An important step toward answering this question, and thereby demystifying experts, is to become deeply aware of the universal continuties between expert and ordinary modes of interpreting the world.

All consent to authority must be understood as necessarily an act of interpretation, however unconsciously it is performed by the agent in question. No one can delegate the initial interpretation by which they assent to the claims of valid knowledge put forward by persons who identify themselves as experts. All make that initial judgment, rationally or otherwise. This generalization is of obvious importance for any analysis of personal responsibility in a technological environment, so let us elaborate a bit.

No two persons can have exactly identical experiences of the world. Interpretations of experience, before becoming operational in social practice, must therefore be negotiated between parties whose cooperation is strategic for the smooth performance of some pattern of practice. This is clearly true even in the case of statutory norms (laws). Here, too, interpretation consists at least in deciding whether a given situation is such as to fit under the criteria of a particular rule such that the rule is applicable to the situation (e.g., not all killing is murder). The necessary lack of fit between rules and the infinity of possible situations is the ontological basis for whatever we may mean sociologically by human freedom. Freedom, in this sociological context of discourse, means that *every* person is an interpreter of the meanings that comprise the social world. Indeed, social control essentially consists in the social and cultural processes through which the fact of every person's moral agency becomes concealed from particular categories of the population and revealed to other sectors. (Such a definition should not be confused with a conspiracy theory of social control. Concealment, like awareness, is in part an aspect of historical, unintended connections between consciousness and society.)

Some societies are organized so as to restrict the distribution of important forms of authoritative agency to particular ruling elites. In other societies all normal members of society are considered "responsible free agents." Even in most of these, however, certain people are designated as more equal than others. Why? Because their mastery of a particular cognitive area of discourse or practice seems to make it socially desirable that they be granted the right, under certain circumstances, to intervene in the freedom of other agents. Such privileged people are normally called "professionals," although not all professionals are equal in the power they are granted through such delegations of authority. Even in the case of the most prestigious professions, the legitimacy of expertise must be negotiated if experts are to be widely accepted as authoritative. (Of course, these negotiations are often concealed from the "lay" public.) These negotiations consist of everything from reasoned discussion to propaganda, to mystification. Where possible, the police power of the state is utilized to render impersonation of experts a crime (e.g., practicing medicine without a license in the United States).

All this results in a polemical atmosphere surrounding experts that must be alleviated by a relatively powerful legitimation of expert knowledge. This is usually accomplished if the meanings that constitute expert knowledge are reified; that is, if they can be made to take on in the lay mind properties of false closure and misplaced concreteness. Hence our tendency to use nouns like "medicine" or "science." This combination (reification on the cognitive level and power on the social level) lends expertise an air of nonnegotiability. This phenomenon is an important requisite for technicism on the level of popular consciousness. Once legitimation of expertise occurs, it is often self-reinforcing through what can be called secondary legitimation. This arises from the sheer placebo effects of expert visibility. (Think of the occasional anxiety generated by the mere presence of a teacher or a judge or the faith-healing effects of secular as well as sacred medicine.)

Implicit in this analysis are some criteria for a *nontechnicist* orientation toward the legitimation of experts. This orientation is one that encourages people to inquire about (1) how the experts' versions of the world came to *evolve* into what they are; (2) the philosophical *bases* of claims to expertise and the counterclaims of significant dissidents; and (3) the *continuities* between expert and nonexpert ways of experiencing the world (i.e., how we all selectively rely on faiths, myths, doubts, assumptions, evidence, and convictions derived from common experience). Such inquiry is a necessary condition for the rational demystification of expertise. Only in this way can reliance on expertise be made to stop short of a total delegation of authority to a technicist imperium of experts. Without such inquiry, no practical philosophy of professionalism and its limits is possible, no distinction between rational and irrational specialization can be set.

In the everyday world, it may be said in summary, we do not normally reflect on the reasons why a person is called "expert." We do not do so for much the same reasons that we do not extensively reflect on exactly what we mean by qualities like goodness, evil, beauty, sanctity, or reasonableness that we attribute to a "good man," an "evil leader," a "beautiful painting," a "sacred place," or a "reasonable argument." Yet all societies have persons whose task it is to reflect on such things. They become protective of their prerogatives. And well they might, for the work accomplished by these masters of legitimation (magicians, shamans, casuists, aestheticians, scientists, jurists) must, as we have seen, also be done to *some* extent by *all* persons as they adapt to the contingencies of everyday life. Seen in this light, the tension between experts and laymen is universal and inevitable.

Why do such tensions not erupt into constant warfare? What contains them so that most societies remain reasonably integral? Those who claim the mantle of expert adapt everywhere to what appear to be three universal conditions for their role legitimation. First, they must remain within certain broad ontological assumptions that dominate the culture of which they are a part. (For example, in our own society, imagine the consequences for a "competent scientist" who insisted on the truth of the Ptolemaic cosmology because it squares with divine attributes better than does the current cosmology.) Second, the experts must respect the range of the mainstream values and interests represented in the population, especially if these are defended by other companies of experts. (For example, in our society many scientists believe in an extreme determinist theory of human behavior. Yet they must function within a sociolegal environment that recognizes the personal responsibility of the self as agent.) Finally, experts must actively seek to placate suspicion on the part of those who suspect them of invading everyman's right to act in the world on his own terms.

I have now reviewed some of the definitional problems involved in discussing legitimacy and expertise in the commonsense context of experience. How have scholars tried to explain the construction of moral intelligibility in some kind of scientific framework? Let us now consider this question.

SOCIOLOGY AND THE SOURCES OF LEGITIMACY

Sociologists have studied the sources of legitimacy in three basic ways. For expedient discourse we may call these legitimacy as convention, legitimacy as intellectual production, and legitimacy as world coherence.[15] I shall discuss each in turn.

LEGITIMACY AS CONVENTION.   By convention, to put it simply, I mean people's propensity to keep up with the Joneses, a propensity interrupted only occasionally by the need to ask why the Joneses do as they do. According to this approach, it may be assumed that if a pattern of behavior is collectively sustained for a long period of time, people simply follow along because that pattern comes to appear as "in the nature of things." This type of argument is officially recognized in those legal systems whose intellectual articulations (statutes) are validated and modified by appeals to "precedent" based partly on customary norms. Also, it is this type of argument that is appealed to by those sociologists who consider power and authority, when all is said and done, to be essentially the same thing. If power is held strongly enough and long enough by a particular group, they argue, authority to hold that power will eventually be imputed to that group by the population at large almost as a kind of natural process (habit, psychic corrections of cognitive dissonance, and so on).[16] This claim is generally pressed with greatest effect against thinkers too intellectualistic in their theories of human nature. Such theories assume that people at large have to have the same standards of rigor, consistency, and justification that intellectuals like to insist upon for themselves.

It is fair to say that stress on convention is an important corrective to any overzealous attitude toward intellectualist criteria of moral intelligibility. But the argument from convention begs many questions. For one thing, how much logical consistency and reasoned justification people need in everyday life is a somewhat neglected issue in social research. Since one cannot simply ask people such a question directly, there are methodological difficulties involved in designing appropriate research. Furthermore, since the power of convention is undoubtedly greater at some periods than others, it is not enough simply to assert its occasional importance. One must be able to show when convention is of primary importance and when it is of only secondary significance. Given the fact that so much of historical scholarship is still the record of elite productions and decisions, it is not surprising that our knowledge of such matters is not more advanced. Finally, the notion of convention by itself does not really provide an explanation of anything. The term hides a host of conceptual problems that still require elucidation in sociology. How, after all, does conventional behavior arise? Most sociologists probably assume a fairly orthodox view of socialization as the basis of convention. Others may have in mind hypotheses stressing concepts like imitation, habituation, or reinforcement. The term convention itself, however, implies for most sociologists a sort of automaticity of behavior. So whatever explanatory account the word connotes, it is likely to be an account that seeks the sources of mechanical, unreflective behavior among humans. For this reason the concept of convention is particularly vulnerable to cooptation for technicist black-box models of society. It is therefore important to keep in mind that behavior outwardly defined as conventional by the scientific observer can conceal diverse subjective orientations on the part of actors. These can range from assent, through reluctant acquiescence, to utilitarian expediency. Here again the potential usefulness of phenomenologically oriented and historically informed social research becomes evident.[17]

LEGITIMACY AS INTELLECTUAL PRODUCTION. There is a view according to which legitimacy is an outcome of systematic articulation, a deliberate production of the intellect. Intellectually formalized "marching orders" obviously have played and continue to play an important role in history. Many studies exist of this process, but they deal largely with elite justifications.[18] Rather few studies exist, however, of such processes among populations not defined as elites and intellectuals; studies, that is, of what could be called the casuistry of everyday life. A program of social research inspired by the work of Alfred Schutz and stimulated by Harold Garfinkel has resulted in some significant contributions to this line of inquiry.[19]

LEGITIMACY AS WORLD COHERENCE. A third strategy in the search for sources of legitimacy rests on the assumption that human environments contain structures of symbolic order often called "world views." World views are thought of as symbolic systems capable of integrating cognitive standards, values, norms, and institutions into patterns of (subjectively experienced) coherence.[20] Some such structures are thought, in Spenglerian fashion, to characterize the boundaries of whole sociocultural periods. On a less grand level, historians concerned with periodization have popularized a vocabulary of coherence structures known to every Western schoolchild (classical era, feudalism, Middle Ages, ancien régime, Renaissance, modernism, etc.).[21] I want to examine this strategy for explaining legitimacy in some detail. Many key problems are posed by it for the scientific nontechnicist study of legitimacy. The existence of a world larger than one's four walls can be discovered by experiencing a sense of radical discontinuity between one's taken for granted ideals and the behavioral routines of one's everyday environment. Some crises of moral intelligibility seem utterly to resist interpretation unless the mind is willing to posit the existence of "deep structures." These are patterns of order remote from the "surface structures," or common sense, rationales employed by ordinary people in negotiating their daily lives. Conspiracy fantasies are often the deep structures of the untutored. (Sometimes, of course, these turn out to be less fanciful than the models of the sociologist.) What, then, are the referents and implications of a term like "world views?" And how could an ordinary person hope to affect such pervasive deep structures?

What should sociologists study when they study deep structures of orderliness? I shall briefly note three lines of inquiry that constitute some steps toward an answer: (1) study of intersubjective assumptions of social practice (issue of consensus); (2) structuralism; and (3) comparative mythology.

If one inquires into the types of consensus among a population about the nature of the world, one can think of the concept "consensus" in two quite different ways.[22] One way is to count heads, add up consciously held attitudes toward things, and then map patterns of agreement. If these commonly held attitudes are very broad and include many dimensions of experience, some would want to say a shared world view exists. (So, for example, a few widely shared attitudes about the nature of God and his power over the world could be said to add up, say, to a Christian world view.)

However, there is a quite different conception of consensus. As against common meanings that are conscious in the minds of participants, there are also intersubjective meanings that are not necessarily consciously available in the subjective lives of most members of a population. Intersubjective meanings are constitutive assumptions that govern patterns of social practice. Any pattern of social practice can be regarded as grounded in beliefs, norms, and values that make that practice intellectually coherent. Consensus on such assumptions is not necessarily a matter of conscious agreement among participants in that social practice. Rather, it is a matter of indirect acquiescence expressed through simple willingness to engage in a given social practice.

One good example is the practice of bargaining in liberal market societies, as contrasted with other forms of negotiation to be found throughout the world.

> Our whole notion of negotiation is bound up . . . with the distinct identity and autonomy of the parties, with the willed nature of their relations; it is a very contractual notion (Taylor, 1971:23).

Also presupposed in a liberal society is a high degree of pressure toward separation between public and private domains of life, with as much of the volume of bargaining relationships left to the private domain as possible. Contracts, that is, are supposed to be free from public interference except for the enforcement of contracts as such.[23] Persons in a liberal society who "bargain in good faith" may not know that they have assumed the validity of certain meanings. But they have. By engaging in the market-oriented sense of bargaining, they have acted out ontological faith in the interest-conscious individual, the moral autonomy of the will, the sanctity of contract, and the importance of utilitarian motives. But these meanings are brought to life in the everyday liberal world by adherence to the *practice* of bargaining as it takes place in such a society. This does not mean that persons do not bring attitudes of all sorts to the negotiation process. What they need not consciously bring into the negotiations, in my example, is the set of ideas and norms constitutive of bargaining itself. Intersubjective meanings, then, are "constitutive of the social matrix in which individuals *find* themselves and act" (ibid.:17; emphasis added).

Another approach that clearly addresses the problem of deep structures of coherence is the sort of structuralism associated with Claude Lévi-Strauss. As Caws says of his work:

> Language, myth, and so on represent the way in which man has been able to grasp the real, and for him they constitute the real; they are not structures of some ineffable reality that lies behind them and from which they are separable. To say the world is intelligible means that it presents itself to the mind of the primitive as a message, to which his language and behavior are an appropriate response—but not a message from elsewhere, simply as a message, as it were, in its own right (Caws, 1970:203).

Structuralism is a method for getting away from Western man's preoccupation with the privileged status of the ego in the world, getting away in the interest of encountering the potentialities of mind through direct confrontation with the world's intelligibilities.

> Just as the assertion that the world is a message elicits the immediate response "from whom?" so the intelligibility of the world seems to be addressed to something more basic and more permanent than the momentary and evanescent subject of particular utterances and particular actions (ibid.:203).

The moral problem posed by this form of structuralism is whether the person as individual agent has been fatally diminished in moral significance along with the person as individual knower.[24]

Other lines of inquiry that are stimulating social science interest because of their relevance to the investigation of deep structures include research into symbolism,[25] root metaphors,[26] and comparative mythology. Because of the possible insights to be gained by regarding the cybernetic vision of world coherence from the perspective of comparative mythology, a note on the concept of myth in social science is in order here.

Myths are essentially explanations of how the world and the society came to be. There are cosmogonic myths, myths of origin pertaining to particular social practices, myths of temporal structure (eternal return cycles, linear progress), and myths of sacred personages (messiahs, folklore saints and saviors, local gods). A very important facet of comparative mythology is the study of eschatologies—the mythologization of time with special reference to the future as destiny. With some notable exceptions, American sociologists have shown little interest in this line of inquiry. This is unfortunate because close study of mythical thought soon reveals the difficulties of venturing a definitive distinction between secular and mythical thought, a point important for the sociology of knowledge. August Comte did not see himself as creating an eschatology when formulating his historical metaphysics of the three stages of progress, nor did Karl Marx with respect to the classless society. With reference to social theories of our own day, Robert Nisbet (1969) has controversially subjected the whole category of "social development" theories (and by implication all alleged laws of social change as such) to the charge that they are not scientific but mythical notions.

Motivational vitality is often grounded in belief in some myth: "Myth assures man that what he is about to do has already been done. In other words, it helps him to overcome doubts as to the result of his undertaking" (Eliade, 1963:141). Perhaps the central mythical theme of the modern West, shared by virtually all secular ideologies, is the image of man as artificer— in some really important way the fashioner of his world and his history. Many events have come to threaten this mythologized self-confidence. Aside from crises like wars, conquests, genocides, and economic disasters, there has also been the evolution of Western civilization itself into a stage of societal complexity that generates simultaneously available and competing world views in the same society.

This cognitive and mythological pluralism is generally celebrated in liberal democratic ideology. But another strain of Western thought has been marked in one way or another by the fear that this condition must end in nihilism; that is, the decline of coherence itself, or the conviction that any sense of coherence is essentially illusory. Such a condition is allegedly indicated by the spread of a subjective sense of moral absurdity, now already amply reflected in the arts. The world comes gradually to be seen as "up for grabs" as there spreads a sense of "anything is possible." If history mediates no truth but is only a record of adventure, then

> it is not inadmissible to think of an epoch . . . not too far distant, when humanity, to ensure its survival, will find itself reduced to desisting from any further "making" of history in the sense in which it began to make it from the creation

of the first empires, will confine itself to repeating prescribed archetypal gestures, and will strive to forget, as meaningless and dangerous, any spontaneous gesture which might entail "historical" consequences (Eliade, 1959:153–154).

Surely the cybernetic pantechnicist model of coherence and order is one mythical fulfillment of such an ending of history!

Must we assume, as some cultists of microsociology seem to, that if deep structures exist, the individual person is helpless before them? We need not assume so. One reason why some people think so is that they believe deep structures of coherence are "causes" of society's surface structures and, through them, of individual behavior. Another reason why some thinkers derive a pessimistic vision of personal helplessness from theories of deep structure is inherent in the phrase "world view" itself. The phrase suggests an object in nature that the individual person metaphorically grasps through the mind's eye. This reliance on the metaphor of vision connotes the passive stance of the spectator who can but gaze out upon the world in visual contemplation. Such pessimism is not in order.

Symbolic coherence patterns are important for the understanding of social structures. But it would be fallacious to think of them as a cause of social structures or even as standing in any clearly definitive "fit" in relation to social structures. Some anthropological observations may be helpful in clarifying this point.

When we examine primitive societies, we find that many of them shared a number of large-scale symbolic assumptions. These included the efficacy of magic in social life, witchcraft beliefs, the importance of the power of ancestors in systems of social control, the absence of any doctrine of death as total spiritual annihilation, some degree of belief in what has been called the High God principle, and various assumptions about modes of healing. Yet these same societies often differed radically in social and political structure. Some were integrated by kinship lineage systems; others had developed structures of political specialization. Some were hierarchical; others, egalitarian. Some had age-grade stratification; others, not. We see from this, then, that a pattern of symbolic coherence can function as constitutive rules for more than one complex of social practices. Much will be made of this point in Chapter 5.

It seems to me clear from what has been said that world view is a misleading term. If we wish to speak of the coherence of a human world as an order comprised of thematically connected meanings, then we cannot speak of this order primarily as a passive "view" or picture. The coherent world is *not*, in this sense, first of all an object of reflection (hence, of reflective knowledge). Coherence, in the sense I am using it, is intimately connected with *action*—with social practice. It is through practice that abstract meanings are concretized in the everyday world. These meanings, then, are not subjective attitudes brought *to* the world. They are already out *in* the world because, as we have seen, they are presupposed in structures of practice.

Why can it not be said that the human world, once constructed, remains legitimate forever, barring great disturbances of external origin? Social thinkers have tried to answer this question by means of various accounts of how delegitimation occurs. I shall briefly glance at some of these efforts now. None of these accounts seems to provide sufficient basis for clarifying why everyone is a moral agent, knowingly and willingly or not. Any effort to support this latter claim would have to rest on a view of delegitimation as not an episodic event but a persisting state. In this view it is episodic events of stable legitimacy that require explanation. Such a strategy must be based on a conceptualization of the human world as unstable by definition. This I seek to provide in Chapter 5.

DELEGITIMATION IN SOCIAL THOUGHT

Social thinkers have long been concerned with the sources of instability in the moral intelligibility of societies. I have just reviewed three standard approaches to explaining the sources of legitimacy: convention, intellectual justifications, and deep structures of symbolic coherence. Most major theories of social change have been based on the converse of these three approaches to explaining legitimacy.

Let me illustrate this generalization by first considering moral *unintelligibility* as a felt *incoherence* in one's world, i.e., contradictions in symbolic coherence patterns. The possibility of incoherence has ever and again inspired in the West philosophers of systematic metaphysics. The current influence of logical positivism and analytic philosophy dilutes the contemporary scholar's empathy with the dependence felt by previous generations of thinkers upon the assurances of order they derived from speculative metaphysics. After all, the philosophical system builders did not think in a vacuum. They were responding to the issues of their day. From Plato, through Augustine, Thomas Aquinas, and Galileo, to Hegel, Marx, Comte, and Spencer, the task was to account for the world—its past, present, and future. Their influence can be noted in the persistent interweaving of their free speculations with the dogmatic ideologies of control and revolution that are so marked a feature of Western institutional history.[27]

When free speculation moves from the philosopher's search for coherence to intellectually crafted legitimations per se in the interests of social control, we get dogmas and laws and casuistries. The converse of such deliberate justifications are *heresy and crime*. These have ever been the province of officials, priestly and secular. The West has been particularly notable among civilizations for the production of dogmas and officials concerned with their enforcement. Perhaps this is so because of the emphasis on rigor and formalization that is so much a feature of Western culture. The

West is notable for a persisting strain toward the formulation of legitimacy in impersonally rational terms, as documented in Weber's great sociological studies. The benefit achieved by such formalizations (the universalization of discourse across parochial boundaries) is ever threatened by the strain toward "legalism." Legalism is one form of imbalance in social practice between means and ends and hence in itself one source of incoherence in symbolic order. Be that as it may, from Socrates' judges to the House Un-American Affairs Committee, the "authorities" have protected against heresy the constructions of official legitimation, and against crime the codified norms of society.

With the advent of modern sociology, the dogmas of church and state were partially replaced by dogmas about society itself, based upon the study of conventional behavior. The converse of conformity to societal dogmas reflective of convention is *deviance*. From Emile Durkheim to Talcott Parsons, sociologists have elaborated on the scientific study of deviance as a source of social change. From "moral statisticians" to contemporary "systems engineers," the agents of social control have tried to measure deviance the better to control it.

In most Western societies all three types of explanation for delegitimation (metaphysical incoherence, heresy, deviance) have come into some disrepute. Positivists, no longer concerned with totalistic coherence, have exiled metaphysicians to academic enclaves in which the latter address almost exclusively each other. Heresy is part of the bill of rights. Officials everywhere are under contemptuous suspicion, having to rely increasingly on power rather than authority. There is a growing concern, as well, that behavior not be indiscriminately subjected to control through the criminal code (e.g., the debate over deregulation of much sexual behavior, of drug use, and of pornography). Finally, the concept of society itself has become so demystified in the years since Durkheim wrote that sociologist-iconoclasts are even now dismantling the last rhetorical justifications for the scientific status of "deviance" as a concept.

The persisting nature of personal agency (and hence of freedom) in the constituting of the human world is obscured for those who uncritically accept explanations of delegitimation such as just reviewed. This is because such explanations make it appear as if delegitimation were just a rectifiable episodic event, a departure from a normal and natural state of moral intelligibility. If moral unintelligibility is taken to be a matter merely of ignorance about some objective feature of the world, one is thus invited to turn to the metaphysicians and ask them to help the agents of social control readjust their formulations of orthodoxy. If moral unintelligibility is thought the result merely of violations of formalized norms, then let us stamp out heresy and crime through inquisition and police surveillance. And if delegitimation is the product of deviance, then bring in the social engineers and the behavior reconditioners. Such is the road to technicism.

Some Transitional Thoughts by Way of a Summary

Robert Heilbroner once depicted in one paragraph the Western common-sense view of the world, which I have taken as necessary for us to reject if we are to sustain the credibility of human dignity.

> To the ordinary person, reared in the tradition of Western empiricism, physical objects usually seem to exist "by themselves" out there in time and space, appearing as disparate clusters of sense data. So, too, social objects appear to most of us as *things*: land, labor, capital; the working class and the employing class; the state and the superstructure of ideas, philosophies and religions—all these categories of reality often present themselves to our consciousness as existing by themselves, with defined boundaries that set them off from other aspects of the social universe. However abstract, they tend to be conceived as distinctly as if they were objects to be picked up and turned over in one's hand (1972:9).

There are no stable objects in nature. Even stones eventually are changed in form under pressure of the elements. Human environments are not stones. Part of what they are is symbolic structures that stretch from the remote world of "once was" through the seeming imperatives of the "now" to images of the "might yet be." All of these times co-exist in the "present," which consists not only of buildings, roads, rules, values and institutions but also of nostalgia, hope, despair, memories, deprivation and desire.[28]

All meanings are abstract until interpreted. For a person to become conscious (on however rudimentary a level) is to "awaken" into this field of meanings. And it is to discover himself an active agent of both the interpretation of meanings and of their practical organization in the everyday world. In this sense, action can be avoided only by dying. To the extent that circumstances include episodes of choice in unpredictable contingent situations—and they always do—to live is indeed to be forced into freedom.[29] This is not meant to deny the facts of socialization. In phenomenological terms, meanings may indeed be encountered by the person as located *within* himself as well as without. But this is a relative matter for empirical assessment, not an automatic assumption that is safe to make with respect to any particular set of meanings. In any case, wherever meanings are encountered, they must be interpreted by the person with reference to situational applications and transactions. Even if "interpreted" is to mean the wholesale acceptance by the person of someone else's interpretations, this, too, is an interpretive act. In other words, meanings should be viewed not as introjected objects but as available patterns of values, norms, and rules. These patterns provide fields of pressure and opportunity for the negotiation of motives, projects, constraints, and legitimations among persons and groups.[30]

Lest my argument be misunderstood, it is well to reiterate this observation: to say every person is a hermeneutical agent is not the same thing as saying that every person is free to constitute the world as he pleases. No one can "name" everything anew. The human environment is already charged with sedimented meanings at the time that any one individual becomes aware of it. Thus, in our society, if we should happen to stumble across a large sum of money, we discover an already "valuable" object and react accordingly. For this to cease to be true, *the whole interpretive horizon* (i.e., the economic and social organization of our society) would have to be altered in some radical way. This is what is so frightening about a phenomenon like "runaway inflation." In a money economy, we experience the instability of currency in the social world much as we would an earthquake in the physical world. When the foundations shake, anything can happen.

The mission of these two chapters on society as a structure of doubt is pursued in the chapter on scarcity. After rejecting the excessively materialistic conception of scarcity still in vogue, I substitute for it a conception that takes account of all sorts of meanings of human "ends" and "resources," both tangible and intangible. A sociologist should not regard "wants and desires" as unique to individuals. Certain sorts of desires are the product of deprivations that are necessary aspects of particular types of social organization.[31] It is true of all forms of social organization that they specialize in generating some sorts of possible desires at the expense of others. Thus, some sorts of profoundly important scarcities, in this larger intangible sense, are a permanent feature of any human environment. In this sense, there is no such thing as a postscarcity world, only the move from some sorts of scarcity to others. The person *becomes* an agent, in a significant sociological sense, by virtue of his struggle to make moral sense of his (and others') actions in a human environment. As such, the person is a "producer" and a "consumer" of interpretations, among them legitimations and delegitimations. Some of these interpretations are articulated in the form of deeds. All of this takes place in a human (symbolic) environment conceivable as a scarcity economy of nostalgia, aspiration, desire, resources, and satisfaction.

Notes

1. I owe the concept of an environment as a "forensic context" to Nelson. A good depiction of what he means by this term occurs in the following quotation:

> I refer to the issue I describe under the heading of *forensic contexts*. Basically in complex societies and cultures of the Western world, all social action has regularly involved reference to bodies of protocols which correlate all notions and evidential canons, associated with the proof or disproof, of arguments for or against any given declaration or claim whether the declaration be about what is or was or ought to be. Among the most important instances of such logics and technologies undergoing development, in my view, are

rationale-systems, structures of reasons, explanations, procedures establishing requirements in respect to truth, virtue, legality, fittingness. Without such rationales, orderly social process and social accounting are unthinkable; the work of the world does not get done. Social and cultural regressions to the so-called "state of nature" manifest themselves when the established rationales go out of phase or lack a compelling and vital center (Nelson and Luhmann, 1976:no pages in text).

2. I trust it is clear that I am not accusing these thinkers of being technicists. Rather, I want to point up the mechanistic nature of the sort of theorizing that lends itself to ultimate usage in a technicist manner by those who would want to drive it that far.

3. I am not concerned here with entering into the debate about what it was that Socrates (or Plato) precisely meant by Socrates' defense. This would entail detailed discussion of whether it is in any way appropriate to regard Socrates/Plato as social contract theorists, and if not, then what. My imagined response of Socrates to Taylor reflects a more modern vocabulary stressing the concept (and connotations) of political culture. Obviously, there are all sorts of ontological considerations that must be added to any account of what Socrates/Plato had in mind regarding virtually anything they said in the context of ancient Greek philosophy. All I would maintain is that, whatever else there was, Socrates' defense reflects at minimum a far richer view of what we would now call the relationship between self and political culture than is present in Taylor's usage of Socrates' defense in his own discussion.

4. This distinction is not, of course, forced upon me by any firm tradition established in the literature. The *Oxford English Dictionary* has over twenty-five different usages each for the terms "state" and "government." Only one usage of the term government (the seventh) contains the distinction I make in my discussion. The case for drawing the distinction I do, however, can easily be defended on both sociological and philosophical grounds. The forms of particular governments clearly diverge, often drastically, from the forms of the states that they supposedly embody. These divergences create serious legitimacy problems. For an example of this point with respect to current United States affairs see Commager's discussion of the "intelligence community's" activities in relation to the U.S. Constitution (1976).

5. This phrase comes from a fine essay by Spiegelberg (1973). The relationship between the individual as individual and the individual as a member of a political "we" is barely touched upon in Taylor's analysis of legitimacy.

6. This position underlies the approach to the concept of legitimacy taken by Gellner. His book is based on the assumption that although there always "tends to be an interdependence between legitimations offered in one sphere and another . . . in our time it is the validation of cognitive practices which is ultimately crucial for other areas" (1974:29).

7. There is a neoprimitivist vein of literature in social science that regards law as the virtual enemy of legitimation by custom, law having been allegedly imposed by elites in the interests of taxation and other modes of social domination. Here the neoprimitivist tradition in social philosophy is drawing on material from the mainstream tradition of nineteenth- and twentieth-century sociology of conflict, which itself did not, of course, end in a neoprimitivist normative stance. See Diamond (1974).

8. This orientation has characterized R. D. Laing's and others' discussions of madness. It also appears in Herbert Marcuse's and Norman O. Brown's notions about the benefits of "polymorphous perversity" and in the mainstream rhetoric of the drug and sex cults.

9. Cf. Schutz (1963). Very much based on Schutz's analysis is Garfinkel's essay on the rational properties of scientific and commonsense activities (1967). The methodological implications of phenomenology and ethnomethodology for research into everyday life have been intensively discussed by Cicourel (1964) and Churchill (1971) among others. It should be understood that the intent among such thinkers is not simply to reproduce the subjective reports of respondents uttered in commonsense terms. This would be nothing but crude ethnography. The intent, rather, is to incorporate models of commonsense experience into the models that the scientific observer generates from a more comparative data base. The problem is to narrow the gap between the universalizing intentions of observer models of society and the particularizing tendencies of participant models, without collapsing one into the other.

10. For example, on the one hand masses of people have unexpectedly capitulated to the wildest propaganda of totalitarian regimes whose demands were a living mockery of the most basic norms and values of Christendom however defined. On the other hand, resistance to these demands occasionally emerges under conditions that are unpredictable from the standpoint of ordinary social science theory. By way of example, consider the virtually incredible account of the life and martyrdom of the Austrian anti-Nazi peasant Franz Jägerstätter, as documented by Zahn (1964). Such stories, in totalitarian and nontotalitarian contexts alike, are not so uncommon as to leave history unaffected. To deal with such matters scientifically, however, it is not sufficient simply to observe behavior. What persons refrain from doing in a situation, and their reasons for inaction, can be as sociologically significant as what they observably do. In the study of legitimacy, insight into such rhythms of action is important.

11. This view is clearly adhered to by Milgram in an experiment that is one of the most significant contributions of experimental social science to moral philosophy in the century (1974). Milgram devotes Chapter 10 of his book to a theoretical interpretation of his findings. His effort is entirely grounded in an evolutionary, biologically reductionist, cybernetic model of species adaptation to the imperatives of complex social organization. He is driven to this approach, one presumes, through having been impressed by his findings about the seeming irrelevance of what people say for what they do in his experiment. However, it should be noted that, as far as the data in the book suggest, Milgram was perhaps so caught by surprise at his findings that he made little effort to probe sufficiently the deeper links among the thoughts, the actions, and the general social experiences of his subjects to come up with a cognitively rich *sociological* interpretation of his results. That remains to be done. As it stands, his theoretical interpretation is a model of the unsatisfactory limits of one sort of cybernetic reasoning applied to human affairs.

12. Regarding this, note the criticisms of Seymour Martin Lipset's approach to legitimacy by Charles Taylor (1971:36 ff.). This is also the position taken by Gellner toward legitimacy and is the conceptual staging area for his attack upon the emotivist position in moral philosophy.

As indicated, the emotivist position is in one important sense far too objectivist—in that it makes itself a present of a given, objective world, which is *then* to be evaluated. . . . The picture is of evaluation floating over objects, as the Holy Spirit moved on the waters, ready to settle down where it chooses. . . . The reason why the idea of moral approval detached in its essence from its objects, sounds so very paradoxical, is precisely this: fundamental moral approval is constitutive of this, that or the other conceptualization (in effect—construction) of the world and thus is welded to the objects it has constructed (1974:41).

13. The existence of lynch mobs, vigilante groups, and racial pogroms would seem to challenge this assertion. But it should be kept in mind that in United States experience, appeal to people's sense of the illegitimacy of these things has proven as important in keeping such phenomena in check as has the exercise of (what would often have proven insufficient) police power. It has, however, frequently been observed about Americans that illegality as a definition of illegitimacy is more likely to inspire consensus than any other way of defining illegitimacy. But it is not so much respect for law as such that accounts for this as perhaps a general sense that in a culturally pluralistic society, the law is the only factor that "holds things together."

14. The Calley case provides an example of what I am talking about. When Calley was convicted of responsibility for the My Lai massacre in Vietnam, polls tended to show that something on the order of eight out of ten Americans thought the sentence imposed on Calley was unjust. Such a statistic tells us virtually nothing. We need to know what was the state of information about the case in respondents' minds? What were the influences of race thinking applied to Vietnamese that affected perceptions of what Calley had done at My Lai? What were the multiple interpretations that were held of the meaning of the verdict itself? And so forth. Without such data we have little way of knowing what the public means by unjust.

15. My friend, Professor Robert Daly, has suggested a rather apt colloquial translation of these categories. An anthropologist visits us from Mars. If he were to think of legitimacy as convention, his question would be: "What's going on around here?" If he thought of legitimacy as intellectual product, he would say: "Take me to your codemaker." And if he thought of legitimacy as world coherence, he would say: "Take me to Max Weber."

16. Since men do not make the just powerful, they make the powerful just, as Pascal said.

17. Schutz has some important things to say about Weber's use of the notion of "traditional" behavior in the course of which he clarifies for the benefit of all sociologists the value of the phenomenological analysis of such concepts (1967:chap. 1). More recently, the works of Pocock are invaluable for those interested in the conceptual clarification of the concept of tradition and likewise in the empirical application to historical sociology of these clarifications (1968, 1973, 1975). There have been a number of technical advances in the methodology for studying conformative behavior, including the analysis of verbal behavior. For example, componential analysis techniques developed in linguistics have been used to study concepts like status and role, disease categories, and values. Much of this literature is gathered in Tyler (1969). Kluckhohn (1961) has used the same technique to examine the nature

and operation of values. There is no scholarly unanimity on the efficacy of techniques of this sort, as some of the essays in Tyler make clear. For some differences among approaches often lumped together, like ethnoscience, ethnomethodology, and phenomenology see Psathas (1968). The sociology of religion has been rather fruitful in empirical studies that bear upon legitimation phenomena. Weber's studies in religion and society remain without peer as examples of how historical materials can be utilized in constructing a sociological account of legitimacy that includes a place for both persistence and change in mass behavioral regularities. The contemporary literature, too, contains some studies notable for their efforts to deal with similar problems. One such neglected example is Underwood's massive field investigation of attitudes and behavior under conflict conditions in an American town (1957). Underwood was concerned with the empirical effects of Roman Catholic versus Protestant notions of legitimate authority and their effects upon behavior. Also useful is Lenski's survey of religious attitudes and behavior in Detroit (1961). Gollin studied two communities of Moravian origin exposed to two radically contrasting national environments: Herrnhut, Germany, and Bethlehem, Pennsylvania (1967). An example of historical sociology in the structural-functional ambience applied to the problem of legitimation is Erikson (1966). A fascinating but neglected study of constructed legitimacy in a formal organization whose bureaucratic hierarchy had no basis for authority is the investigation of the American Baptist Convention by Harrison (1959). In a related vein is my own study of the efforts of the organizational elites of the Jehovah's Witness movement to maintain their movement in a hostile environment between 1880 and 1960 (Stanley, 1969). I have cited these various studies because despite their specialist variety they all bear upon legitimacy as a central intellectual problem. It is unlikely that these studies would show up under the term legitimacy in any bibliographical classification scheme, which is one reason for dubiousness about the value of such schemes.

18. For the concept of casuistry see Kirk (1927). For a study of the context in which "conscience" becomes transmuted into "interests" in liberal society see Gunn (1969). For a doctrinal history of the legitimation of political acts by "reason of state" see Meinecke (1957). For two studies focusing on individual thinkers who played an important role as intellectual brokers in adapting philosophical systems to the ideological-moral problems of their times see Krieger (1965) and Richter (1964). The latter work is a study of T. H. Green's fusion of Hegelianism and pre–World War I British liberal philosophies of political obligation. Also relevant to the study of intellectual articulations of doctrine in the struggle over legitimation are Nelson's study of the long conflict over the meaning of the deuteronomic condemnation of usury among brothers (1969) and Macpherson's study of the political theory of possessive individualism (1962). In this work Macpherson argues that the "market society" was legitimated through the influence of seventeenth-century political theorists on the intellectual level who were themselves articulating the doctrinal significance of social organizational changes that had been forming in England for many generations. Historical studies of legitimacy crises always illustrate the dependence of events upon the deliberate fusing in real life of what has been called here legitimation by reference to intellectual constructions, legitimation by appeal to conformity, and legitimation by reference to coherence. See, for example, de Santillana (1955) and Hanke (1959). The investigation by de Santillana of the

processes leading to Galileo's condemnation shows in many ways how things could have gone differently, depending always upon how interpretations of existing meanings were made at particular points. Galileo, for example, largely remained within the issues of theological interpretation, according to de Santillana, in defending the orthodoxy of his work. Yet the author shows how significantly other considerations bore upon the case. Among these were Pope Urban VII's strategic position in European politics; the struggle for control of elite education by the Jesuit order; the public relations problems of a thought police (the Holy Office of the Inquisition); and the anxiety of some theologians over any possible intellectual "scandal" such as the conceivable unreliability of the Aristotelian substructure of Catholic dogma. Yet de Santillana does not neglect the intrinsic logic of the neo-Aristotelian principles of coherence itself such as, for example, the sharp distinction between the earthly and the heavenly realm that Galileo's philosophy of nature seemed to be contradicting. Hanke's study describes the great debate between Sepulveda and Las Casas in 1550–1551 over the applicability of Aristotle's theory of natural slavery to the indigenous peoples of Latin America. This debate was an official struggle between interpretations that, Hanke shows, was to have significant influence upon Spanish imperial policy for a long time.

It may be wondered why subsequent generations honor great intellectuals who contribute to the legitimation of some institution or policy. Reading historical studies of such efforts suggests at least one answer. Perhaps such men are honored because their efforts demonstrate personal potency in the face of situational challenges. The intellectual productions of significant intellectual ideologues do not threaten with mockery the persistent human striving after universal meanings that fulfill men's desire to experience history or nature as morally intelligible whatever their own momentary interests. At the same time, these productions demonstrate how universally significant meanings can be made potently relevant to situational demands. Our present debunking attitude toward legitimation ideologies should not be allowed to obscure their significance as demonstrations of world construction.

19. Apart from the works of Schutz see Mandelbaum (1955), Garfinkel (1967), Cicourel (1964, 1968, 1974), and Jack D. Douglas (1967, 1970a, 1970b). The literature of ethnomethodology is important because it bears fundamentally upon the active, freedom-using capacities of human agents. This point sometimes needs to be salvaged from the rather cultic polemics toward which people writing in this vein incline.

20. The term "coherent" should not be taken in this context with logical literalness. It is meant, rather, in the nonlogical, pragmatic sense one has in mind when one thinks of the singularity of a taken for granted world. A number of major sociological thinkers are associated with efforts to inquire into the structures of such unities. Most especially the names of Wilhelm Dilthey, Georg Simmel, and Pitirim Sorokin come to mind. Among historians, Jacob Burckhardt, Marc Bloch, and J. Huizinga are noted for their efforts to define entire historical periods in such integral terms. And one also thinks, with respect to architecture, of the efforts of Henry Adams and Erwin Panofsky to analyze the integral cultural unity of the Gothic cathedral. A landmark methodological contribution to such questions as they appear in sociological theory is Brown (1977).

21. For useful approaches to problems of defining order on this scale see Kuntz

(1968), the works of Eric Voegelin, and those of F. S. C. Northrop. There have been some imaginative empirical studies from various directions all organized around an interest in the concrete manifestations of coherence principles of order. For example, Curtin traces the impact of world view themes (e.g., evangelicalism, "natural man" theories, race doctrines, evolution, and even specific physical theories like phlogiston) upon the manner in which data from West Africa were received and interpreted by various social elites in Great Britain between 1780 and 1850 (1964). Curtin is able to account for many aspects of the subsequent style of British imperialism in Africa and its legitimation. Kendrick was interested in the influence of world view themes upon specific reactions to the Lisbon earthquake disaster of 1755 (1955). The point of citing these few examples is to show that sociologists of knowledge interested in the concept of legitimacy can ill afford to ignore the relevant contributions of historians with similar interests. Finally, one should note the existence of one of the very few efforts to study a major public policy question from the standpoint of ideas that inform this chapter. Geiger attempted to apply a number of social theories bearing upon world views and social change to the task of understanding the failures and frustrations of the American foreign aid program (1967).

22. I owe my appreciation of the distinction between these two senses of consensus to Charles Taylor (1971). I have adopted and expanded somewhat his distinction between "subjective meanings" and "intersubjective meanings."

23. Indeed, one of the contradictions in liberal capitalist societies today is that interest-bargaining units have grown in size so far beyond "human scale" (i.e., personal scale) that the original assumptions that legitimated the distinction between public and private life have become virtually irrelevant. It is this realization that is expressed in phrases like "private government" and "corporate socialism." Contradictions of this sort clearly threaten the coherence of the liberal order.

24. There is no way to determine this without integrating into relevant social science discussions perspectives from the philosophy of action. An example of a step in this direction by a philosopher is Bernstein (1971). I hope this book is another. But very much more remains to be done.

25. One of the more important approaches to symbolism has been the focus on representative symbols. This designation refers to the symbols—linguistic and other—by and through which members of any collectivity try to accomplish the following: define and experience themselves as communally bonded to each other, distinguish between legitimate and illegitimate norms and actions, differentiate themselves from other communities, and relate themselves to some myth of transcendence. See Voegelin (1952), Warner (1961), Spencer (1969), and Mary Douglas (1970), among others.

26. For the concept of root metaphors see Pepper (1970). According to Pepper, root metaphors are really fundamental hypotheses about the world's basic metaphysical structure. Root metaphors begin as analogues from commonsense experience. They are elaborated by geniuses who see their philosophical possibilities for explanatory integration of observations and are eventually systematized in the form of what Pepper calls "world hypotheses." Aside from animism and mysticism, which he considers based on inferior metaphors, four fundamental Western root metaphors are dealt with at length. Pepper tries to demonstrate their bases in common sense, their metaphysical stages of construction, and their implied theories of

truth. These metaphors are formism (Platonism), mechanism, organicism, and contextualism (pragmatism). The literature subsequent to Pepper's 1942 volume strongly encourages a view that the study of sociocultural influences of metaphors is proving to be a most effective way in which the data of the humanities can be brought to bear upon a phenomenologically oriented social science. For a masterful demonstration of this generalization see Brown (1977).

27. On the interplay between theory and practice on the conceptual level see Lobkowicz (1967). For the general sociology of ideology in Western history see Mannheim (1936) and Gouldner (1965, 1970, 1973, 1976).

28. I am speaking of the concept of environment in Chapters 4 and 5 in a somewhat singular manner—as a field of meanings. Actually, it is useful to think of the person as surrounded by three analytically separate kinds of environment that can be ranged along a continuum of "thickening facticity," i.e., increasing resistance to symbolic reinterpretation. Environment I, then, is the realm of cultural symbols per se (art, religion, language, law, etc.). This domain is most open to symbolic reinterpretation and manipulation. Environment II is the realm of social structures per se, studied by the formal sociologist. This domain is somewhat less open to symbolic manipulation because it is dominated by formal structural regularities, amenable to analysis by systems and organization theory, sociometry, network analysis, and other such methodologies of inquiry. Environment III is the realm of heaviest facticity because it is dominated by the physical laws of nature. Simply stated, you can reinterpret the gods to the point of nonexistence in a way that you cannot with a tree or a hurricane. It is still the case, however, that in every society all three environments are embedded in macrophenomenologies of order; superordinate principles of interpretation that link all three environments into some semblance of a coherent world. As societies evolve into complex civilizations, these macroprinciples of order are put under increasing strain, threatening the world with fragmentation and eventual incoherence. Contemporary Western civilization has probably advanced further along this road than any other in history.

29. The whole problem of choice from a phenomenological point of view is very complex. For some of the issues see Schutz (1967:66–69).

30. For a view of motives along these lines see Mills (1940); Blum and McHugh (1971), and Scott and Lyman (1968).

31. Naturally, one can speak this way only on the level of ideal-typical analysis. Real-world societies are never pure exemplars of singular forms of organization.

5 The Structures of Doubt

SCARCITY AND THE ECONOMIES OF DESIRE

What is it that persistently destabilizes the world so as to make of it a permanently forensic context? This, it will be recalled was the second major question stipulated in the introduction to Part II as having to be answered if a philosophy of dignity is to be translated into a sociology of authorship.

The answer I want to pose in this chapter is a conception of permanent scarcity endemic in the human condition to which all persons must relate. Accordingly, this chapter is about scarcity and how to conceptualize it. I begin with a critique of approaches to scarcity that envision its "conquest." Victory over scarcity has been put forward by some as the solution to the problem of technicism. For such a "solution" we will wait forever.

Subsequent to this first section, I move directly into the question of what scarcity is; what its relationships are to patterns of large-scale social organization and to the vicissitudes of subjective desire. The emphasis throughout is on the relevance of these theoretical reflections to a nontechnicist sociology of persons and action.

The Conquest of Scarcity

In some quarters, the decline of social controls based on metaphysics, dogmas, and theories of deviance has inspired some heady antinomian visions of liberation from virtually all institutions of social control. For some, this liberation will be marked by an ecstatic transvaluation ("polymorphous perversity," etc.) of most societally sponsored distinctions between legitimacy and iniquity. Underlying this millennial hope is a notion of limitlessly

117

applied physical technology. When this notion is present, the millennialism of "liberation" takes the form of a theory about the "conquest of scarcity." Postscarcity millennialism is as yet actually more of a theme than a theory, a will-o'-the-wisp of hope than a formula of the intellect. But it hovers between the written lines of some ideologues and as a sort of mental foam around the lips of a few Dionysian alchemists who appear to feel that a culture of psychotropic drugs and cybernetic technology will render obsolete the hard problems of social organization and moral reasoning. The conquest of scarcity is a myth of the fullest eschatological stature. This is obvious once one recognizes that there are really only two final solutions to the ubiquitous problem of human desires. One is the Oriental tradition, the striving toward release from the wheel of wanting. The other is the Occidental way, the bending of nature itself to the demands of human will. What Nirvana means to the ego that would escape all fleshly desires, the conquest of scarcity means to the ego that would fulfill all such desires. In one, all is allowed because all is One. In the other, all is allowed because anything is possible.

Not all end-of-scarcity discussions are informed by a millennial vision. There are at least three versions of end-of-scarcity discourse that I shall now briefly discuss in ascending order of their eschatological significance.

One version of the conquest of scarcity thesis is based on the argument that although natural scarcities are a constant condition of human history,

> what was new, from the seventeenth century onwards, was the prevalence of the assumption that unlimited desire was rational and morally acceptable. When this assumption is made, the real task of man becomes the overcoming of scarcity in relation to infinite desire (Macpherson, 1973:18).

So long as this (capitalist) vision of human nature persists, there can be no end of scarcity because in such a culture "desires will multiply as fast as technological advance can meet them, so that society as a whole will have reason to work as hard as ever" (ibid.:20). It is only on this assumption of man as an infinite consumer of utilities that scarcity must be regarded as permanent.

> No increase in productivity, however great, will end scarcity while people continue to see themselves as infinite consumers. A comparatively modest increase in productivity, or no increase at all in the present productive capacity of the economically most advanced nations, would end scarcity if people came to see themselves (as the justifying theory of liberal democracy must assume them to be) as doers, exerters, enjoyers of essentially human capacities. . . . In such a liberal-democratic theory, the standard of material wants from which scarcity is to be measured is the amount of material goods required to enable everybody to use and develop his human capacities (rational, moral, aesthetic, emotional, and productive in the broadest sense). This bears no assignable relation to the amount needed to meet the supposed or projected actual wants of men culturally conditioned to think of themselves as infinite consumers (ibid.:61, 62).[1]

Critics of scarcity envisioned in this manner might well label life in such a society the "politics of anomie." Their critique is directed against the stimulation of an unregulated infinity of desire by a socially sponsored ideology of utility maximization.

Another version of the conquest of scarcity thesis is based on the argument that present experiences of scarcity are essentially the result of the artificial suppression of a high technology now capable of turning many scarce goods into free or quasi-free goods to the benefit of all people. This approach underlies the "politics of abundance" movement, which sees in unlimited material abundance the telos of industrial society. (Exemplary here are the views of Robert Theobald, 1961 and 1965.)

Some who might accept a moderate politics of abundance stop short of a third perspective that carries the end of scarcity theme to the point of an eschatological vision: the secular millennium of anarchist liberation from all authoritative hierarchy. It is this perspective that generates a "politics of liberation."

> To view the word "post-scarcity" simply as meaning a large quantity of socially available goods would be as absurd as to regard a living organism simply as a large quantity of chemicals. . . . Post-scarcity society . . . is the fulfillment of the social and cultural potentialities latent in a technology of abundance (Bookchin, 1971:11).

What are these latent potentialities?

> A qualitatively new situation emerges when man is faced with a transformation from a repressive class society, based on material scarcity, into a liberatory classless society, based on material abundance. From the decomposing traditional class structure a new human type is created . . . the revolutionary. This revolutionary begins to challenge not only the economic and political premises of hierarchical society, but hierarchy as such.
>
> . . . Anarchism is a libidinal movement of humanity against coercion in any form, reaching back in time to the very emergence of propertied society, class rule and the state. . . . The anarchic movements of the past failed largely because material scarcity, a function of the low level of technology, vitiated an organic harmonization of human interests (ibid.:190, 211, 212).

In this eschatology, liberation from scarcity is accompanied by retribalization of society.[2] Anarcho-communist organizations

> are built around intimate groups of brothers and sisters—affinity groups— whose ability to act in common is based on initiative, on convictions freely arrived at, and on a deep personal involvement, not around a bureaucratic apparatus fleshed out by a docile membership and manipulated from above by a handful of leaders (ibid.:214–215).

As even this simple classification suggests, anyone presuming to enter the debate about scarcity and its role in history must encounter a bewildering diversity of polemic (scientific or ideological), directed as often at straw men as at real issues.[3] Yet we cannot avoid it. No discussion of legitimation as a necessary dimension of personal agency (and hence of dignity) is possible without the help of a comprehensive conceptualization of scarcity.[4] A millennial hope for some definitive conquest of scarcity is an impossible resolution of the technicist challenge to a philosophy of persons. There is no salvational technology that can turn every object of desire into a free good. Some objects of desire are intangibles of the heart that no technology can produce. To understand more clearly what this means, and how it makes persons of us all, we must see in what sense it may be said that scarcity is a permanent feature of the human condition.

The Concept of Scarcity

Any sophisticated discussion of scarcity in human destiny should, if it hopes to avoid the pitfalls of past confusions, take off from at least four generalizations that strike me as reasonable conclusions to be drawn from previous debates.

First, we should not identify as natural scarcities phenomena that are the product of institutional arrangements. For example, capitalist ideology has long assumed a virtual natural scarcity in the motive to work, an assumption that has strongly influenced a variety of public welfare policies. Yet anthropological evidence shows that no such natural scarcity of work motive exists, attitudes toward work being dominated universally by institutional arrangements.[5]

Second, we should not assume things to be scarce that may not, in fact, be. A major example will clarify how this possibility can arise.[6] Consider the notion of personal energy scarcity when the idea of human energy is derived metaphorically from our understanding of physical energy. When it is not recognized that terminology about human traits such as "energetic," "forceful," and "aggressive," are in essence metaphorical, we tend to theorize about human traits in ways that reflect our experience with the physical models that serve as the basis of the metaphors. Uncritical control of discourse by metaphor thus may blind us to the observation that persons do not, for example, necessarily suffer energy "depletion" when engaged in some activities but may experience energy "renewal." This observation leads to a very different view of the economics of activity from that growing out of the assumption that a conservation of energy law applies to human activities such as to require of the ego a zero-sum game policy of energy "allocation."

Third, we should not restrict our concept of scarcity to merely tangible things. Since this theme will be elaborated at length subsequently, there is no need for examples here.

Fourth, we should not oversubjectivize (in the sense of totally individualizing) the notion of values and ends such that scarcity analysis becomes equally subjectivized. What is at stake here may be clarified by considering some passages from Lionel Robbins.

> So long as the theory of value was expounded in terms of costs, it was possible to regard the subject-matter of Economics as something social and collective, and to discuss price relationships simply as market phenomena. With the realization that these market phenomena were, in fact, dependent on the interplay of individual choice, and that the very social phenomena in terms of which they were explained—costs—were in the last analysis the reflex of individual choice—the valuation of alternative opportunities . . . —this approach becomes less and less convenient. The work of the mathematical economists in this respect only sets out particularly boldly a procedure which is really common to all modern theory.
>
>   . . . From the point of view of pure Economics [economic goods or disutilities] are conditioned on the one side by individual valuations, and on the other by the technical facts of a given situation. And both individual valuations and technical facts are outside the sphere of economic uniformity. From the point of view of economic analysis, these things constitute the *irrational* element in our universe of discourse. . . . A moment's reflection should make it plain that we are here entering upon a field of investigation *where there is no reason to suppose that uniformities are to be discovered.* The "causes" which bring it about that the ultimate valuations prevailing at any moment are what they are, are heterogeneous in nature: there is no ground for supposing that the resultant effects should exhibit significant uniformity over time and space (1946:69, 106, 107; emphasis in original).

Many commentators, vulgar and sophisticated, have entered the lists against this account of what economics should be about.[7] (For one of the best see Walter Weisskopf, 1955 and 1971.) Not all critics have kept in mind what Robbins was and was not talking about when he formulated his arguments. For example, Robbins's criterion of economic "uniformities" is measurability over time of the sort capable of yielding valid scientific laws. What he was denying was our ability to give "numerical values to the scales of valuation, to establish quantitative laws of demand and supply" (1946:107). In this Robbins is surely in agreement with many of his more value-absolutist critics.

This caveat, however, does not dismiss the issue in the context of our present discussion. What is at stake here is whether the ends that underlie the experiences of scarcity are in a state of what could be called chaotic subjectivity. Is it really the case that nothing can be said about the sources of scarcity except that they are the collective vectors of random subjective wants of individuals plus the technical conditions of things in their status as resources? Why would such a view disturb so many thinkers?

The answer is that, if taken literally, Robbins's remarks imply a vision of society comprised of individuals who share virtually nothing except that they have wants. Furthermore, these wants cannot be spoken of from the economic standpoint as rational or irrational. Wants simply are. Yet economics is supposedly the science of rational choice. How are we to speak of reason, then, and of value? Are we restricted to the technological reason of efficient relationships between technical means and randomly held ends? If subjective wants are ground zero of the moral universe, the starting point of discussion about rational decision, what are we to make of that aspect of the individual that is comprised of the introjected elements of the moral norms, values, aspirations, ideals, symbols, institutions, social structures, and identities that existed prior to any given individual, into whose midst he was born, and that will persist after his death? What are the implications of saying that we are here entering upon a field in which no uniformities are to be discovered just because we cannot give mathematical expression to them? Surely such a criterion is too strong. There are other grounds for discussing uniformities. These grounds are the socialized nature of the individual, as well as the persisting social structures, cultural standards, and communities that mediate between shared identities and subjective wants.

Individual wants are not, after all, so radically subjective. Social and cultural environments "encourage" persons to have some wants rather than others. There are some objective uniformities about these environments. Such uniformities allow us to speak of objective scarcities implicit in the structures of a given social and cultural order. They allow us at least loosely to distinguish between (objectively rooted) scarcities and (subjectively felt) deprivations. And these uniformities allow us, finally, perhaps even to make statements about wants that are more or less rational, more or less based on insight into self and society, and more or less capable of fulfillment.[8]

With such generalizations in mind, how may we proceed? It will help to reiterate the purpose of these reflections. My concern is not with a general theory of scarcity. The topic is the nature of a truly human environment, the symbolic world in which persons live and act. I am inquiring whether or not such an environment is necessarily permeable to the actions of all persons in their capacity, at least, as hermeneutical agents and if so, how? The answer that I am working toward is that human environments are places of tenuous solidity, forever needing to be reinterpreted, forever generating doubts and aspirations. What is the major generator of this instability of meaning and desire? It is a permanent condition of scarcity; scarcity not of any one thing *but of stability and closure of meaning itself.* Without such scarcity, I do not see how there could be freedom of persons, dignity, and progress.

To make this argument, I first had to clear the ground of other ways of speaking about scarcity. These other ways are not necessarily different in essence from the way in which I shall speak of scarcity. Rather, thinkers who discuss scarcity generally do so incidental to the context of some larger intellectual project.[9] It is this larger project that influences (if not determines) how scarcity is defined and used as a concept. This is perfectly legitimate. But it clears the air to locate one's own usage in the spectrum of other people's usages. To proceed, then, how may scarcity be spoken of in relation to an environment permeable to the interpretations and actions of persons who reside in it?

Scarcity may be defined most generally for present purposes as a shortage of resources regarded as necessary to implement any given end or intention in the form of action appropriate to the demands of concrete situations. An assertion that some such resource is scarce can be made by an "expert" observer on the basis of an analysis derived from a theory about the relevant phenomena. The assertion of scarcity can also be made by someone who—as a participant in his society—applies his rules of commonsense interpretation to some subjectively felt experience of deprivation that appears to him the result of an objective "lack" in his environment. It should be clear that the social types I call "observer" and "participant" are analytic extremes. No social science commands any theory that is so pure in its comprehensiveness and separation from the categories of a parochial commonsense system as to make of the social scientist a pure observer. Likewise, all commonsense systems of thought have some theory content of an abstract, propositional nature that the participant checks against his comparative experience. Such a process makes him in part an observer, on however rudimentary a level.

My interest in formulating an adequate concept of scarcity will not be accompanied by any attempt to develop a comprehensive theory of scarcity. A full-fledged theory of social change as the product of collective and persisting strains toward delegitimation would have to address three general issues. First, in what specific ways can a given society (viewed as a structure of meanings) be thought of as a field of *potential scarcities*? Second, under what conditions are such potentialities selectively concretized into *experienced patterns of deprivation* among particular sectors of a population? Third, under what conditions are these experienced deprivations linked to remedial collective *action*?

These three general questions can be asked with regard to various levels of reference that any comprehensive analysis of scarcity applied to historical cases would have to take into account. These levels of reference are the phenomenological, the sociological, the social psychological, the ideological, and the collective action aspects of existence. A brief word on each will be helpful prior to stating the relevance of what has been said to my larger concerns.

By the *phenomenological* level I refer to the fact that all meanings can be analyzed dialectically as to their potential implications.[10] Any idea's range of implications is neither singular nor infinite in scope. Dialectical analysis can be helpful, as I shall later attempt to illustrate, in laying open potential ways in which any given meaning could come concretely to be experienced in everyday life.

By the *sociological* level I mean the patterns of concretization (rules, roles, institutions, associations, and their structural effects) according to which meanings have become practically organized in any given society.

By the *social psychological* level I mean the social distribution of relative deprivation experiences in the context of a given society.

By the *ideological* level I mean those systematized accounts that, in the form of ideologies produced by intelligentsias, diagnose the moral state of the status quo. Such ideological formulations may or may not include recommended programs of action.

By the *collective action* (praxis) level I mean those organized patterns of action, embodied in social movements, that evolve to rectify some ideologically diagnosed experience of deprivation.

I intend to show how dialectical analysis of existing patterns of social organization can reveal objective scarcities that underlie potential experiences of deprivation. The abstract scarcities can be thought of as potential generators of trends toward delegitimation of the existing social world. These same abstract scarcities are also the foundations of the interpretative *reconstructions* of the social world that—translated into social action—point toward social and cultural change. *All persons are moral agents because all persons take part in the interpretational processes that result in the five analytically defined levels of reference set forth above.* This is true even of the person who, through ignorance or other reasons, delegates his agency as interpreter of meanings to others (such as "experts").

At this point a detailed example would be helpful of how the social scientist (and the moral philosopher) may proceed in detecting the potential scarcities inherent in the plenitude of implications that comprise sociocultural meanings.

IMPLICIT SCARCITIES IN SOME MODES OF SOCIAL INTEGRATION

For purposes of this example I focus on certain abstract models of socioeconomic integration that have been developed by scholars interested in intersocietal comparisons.[11] Specifically, I adopt here the typology developed by Karl Polanyi and his co-workers, who were building on the work of Marx, Malinowski, Mauss, and others.[12]

The typology distinguishes among three basic types of transactional structure of a scope capable of integrating whole economies. They are "reciprocity," "redistribution," and "market exchange." These terms refer to different structural conditions surrounding social transactions. These terms also refer to differing patterns of institutionalized meanings in terms of which such transactions are phenomenologically perceived by participants in the various social systems. (For example, "bride price" is no longer considered by observers, as it once tended to be, to be an economic act of "buying a wife.")[13]

> Reciprocity denotes movements between correlative points of symmetrical groupings; redistribution designates appropriational movements toward a center and out of it again; exchange refers here to vice-versa movements taking place as between "hands" under a market system. Reciprocity, then, assumes for a background symmetrically arranged groupings; redistribution is dependent upon the presence of some measure of centricity in the group; exchange in order to produce integration requires a system of price-making markets (Polanyi, 1957:250).

It must be understood that this is not psychological but institutional analysis. For example,

> Reciprocity behavior between individuals integrates the economy only if symmetrically organized structures, such as a symmetrical system of kinship groups, are given. But a kinship system never arises as the result of mere reciprocating behavior on the personal level (ibid.:251).

Symmetries can form along many social axes. Kinship formations, neighborhood settings, totem groups, and city-states are historical instances of social formations that because of their "human scale" facilitate the growth of symmetrical linkages. Within them there can form any number of voluntary symmetrical linkages of a military, vocational, religious, or social character. Likewise, social systems can have multiple appropriational and redistributive channels (e.g., special patronage groups, family based insurance mechanisms, or voluntary welfare associations). Furthermore, societies can contain multiple major loci for price and power bargaining, as was suggested some years ago by John Kenneth Galbraith in his theory of countervailing power in market exchange societies (1952). Finally, any society can contain all three modes of integration on various levels of subordination. For Polanyi,

> Dominance of a form of integration is . . . identified with the degree to which it comprises land and labor in society. So-called savage society is characterized by the integration of land and labor into the economy by way of the ties of kinship. In feudal society the ties of fealty determine the fate of land and the labor that goes with it. In the floodwater empires land was largely distributed and sometimes redistributed by temple or palace, and so was labor, at least in its dependent form. The rise of the market to a ruling force in the economy can be traced by noting the extent to which land and food were mobilized through exchange, and labor was turned into a commodity free to be purchased in the market (1957:250).

This emphasis on land and labor is not arbitrary. Land, after all, is in most societies the major manifestation of the physical world and its constraints.

Labor is most men's predominant way of expressing their active agency in the world. More than anything else, it is how land and labor are institutionally defined that objectively determines how members of a society will experience the meanings of their relationships to nature and to their own productive powers.

I will use this typology to illustrate the notion of scarcity by implication. Consider the following question: what sorts of experiences and frustrations would be induced by each one of these modes of integration were it alone to function as a single, completely dominant mode of societal integration? (We may ignore, for present purposes, the fact that such singular domination hardly ever occurs in real societies.) It will be argued that reciprocity is logically and empirically central to that complex of values and norms historically associated with the notion of "community." Redistribution is likewise associated with "social welfare" mechanisms and goals, whereas market exchange is identified with "innovation" generated by efforts to respond to consumer demands. Let us explore these associations in more detail.

RECIPROCITY.   The social structures characterized by reciprocity as a dominant mode of integration have been associated ideologically with what has been called human scale. This is because such structures allow for the symmetry of long-term interpersonal commitment, obligation, and growth of emotional identification. These themes are included among what has long been meant by community. Since the time of classical Greek philosophy, the concept of community has been associated with arguments for limitation in group size.

Despite these benign moral themes, however, sentiment has clearly become ambivalent about communities of reciprocity in the West for at least two basic reasons. First, the continuity of such structures has in the past largely rested on conformity to social statuses into which one is born. This has been experienced as restrictive of personal initiative, ambition, and freedom of choice. Second, such communal systems have also been associated with a certain lack of adaptability to social evolution. This is because such structures resist growth in organizational scale, complex division of labor, and other secular, technological motives. Hence the frequent recommendations from social scientists that traditional cultures must be undermined if "economic development" is to take place.

REDISTRIBUTION.   As a mode of integration, redistribution is associated with a tendency toward the centralization of power. It is one important basis for efficiency of resource allocation and, if the central authority is strong enough, for minimization of what would otherwise be endless conflicts of interest between diversified veto groups.

However, redistribution is also associated with the development of autocracy. Furthermore, a society integrated exclusively by redistribution can produce an alienated and dependent population unaccustomed to participation in basic decisionmaking and with no sense of responsibility for the management of societal processes.[14] Finally, as social critics in both socialist and capitalist societies have been pointing out, there is the possibility in redistributive systems for exploitative social stratification through appropriation of resources by a privileged redistributive bureaucracy. Thus, redistribution implies a scarcity of opportunities to exercise personal skills relative to "grass roots" participation, moral responsibility, and some sorts of individual efficiency.

MARKET EXCHANGE.   Sociologically, the history of the utilitarian market exchange principle has been associated with intensive technological-economic development via the appropriation of physical nature and of labor as factors of production and the mobilization of competitive energy for life in a society viewed as a commodity market. However, the institutionalization of the market economy has also been associated with events regarded as subversive of many humanistic values, both sacred and secular. Such events include social class formation based on accumulation of great wealth through market operations; social and ideological rigidities, in the face of human suffering, because of indifference to change for any reasons other than greater market advantage; and the gradual transformation of all forms of value into the commodity criterion of marketability.

Furthermore, it has been argued by some social critics that movements ranging from fascism through communism to extreme nationalism can be viewed in part as facilitating a kind of "retribalization" of society. According to such arguments, totalitarianism should be seen as providing a mythologized equivalent of the communalism associated with the structures of reciprocity. Thinkers holding such views assume that something as apocalyptic as totalitarianism can be explained only as the effect of a most extreme kind of deprivation—namely, deprivation of opportunities for the day-to-day expression of communal values and actions. Such a degree of deprivation must be closely connected with some central feature of the liberal market exchange principle of social integration. This central feature of market liberalism has been the notion of society as an opportunity structure for the fulfillment of an egoistic individualism. According to the critics, however, this notion of society has helped bring about the anomic reduction of community to theoretically atomistic egos related to each other only by the competitive struggle for life chances on the commodity market. The artistic, philosophic, and scientific literature centering on concepts like alienation and anomie has long been implying a causal connection between the market principle of social integration and scarcities of opportunity to concretize the values and norms of community. The literature of social planning from August Comte to Karl Mannheim and the current rhetoric of community development and the new towns movement obviously also bear upon these kinds of scarcities. And, it is hardly necessary to point to the monastic communities of the contemporary counterculture.

Given such instabilities in experience, the importance of nostalgia in generating social movements should come as no surprise. It is strange that the topic of nostalgia has not received more attention from sociologists. Nostalgia may usefully be defined as the perception of the past in terms of the phenomenology of present scarcities. Thus, the atomized member of the lonely crowd, descendant of the rebel who spurned conformity to the communities of reciprocity because of his desire for new horizons, now may look back upon the image of community with romanticized longing. But it is a longing born not of historical experience with the object of his desire, about which he may know very little. Rather, the longing is the mirror image of the frustrations of his own time. Such nostalgia can find expression in forms ranging from popular art fads all the way to political and cultural movements of romantic reaction (a notable feature of nineteenth-century European antibourgeois conservatism).[15] One should note, too, that nostalgia need not fasten only upon a romantically redeemed past (retrospective nostalgia). It can also fasten upon a mythologized version of the original dream of liberation from the past associated with the founding ancestors of the present. In this context, the present is turned into a result of a dream betrayed, and the nostalgia is pitched toward some past moment of eschatological faith in the future (prospective nostalgia). Such experiences can generate social movements of charismatic or revolutionary renewal that reject the betrayed present and seek to restore the original vision of transition from past to future. The future becomes the time of the dream redeemed.

Recapitulation and Conclusion

A bit of recapitulation should be helpful at this point. Most meanings, I have argued, are standardized according to some operational connection with specific objects or situations. Life consists of objects and situations whose range extends far beyond those that lend meanings their initially standardized criteria of interpretation. Thus, many, if not most, meanings are vulnerable to the constant pressure of new experience. Of course, meanings are not infinitely malleable. They have an immanent logic of conceptualization, and a host of assorted empirical associations, that gives them their intellectual coherence and integrity in the social world of action. Yet most meanings do carry implications beyond those that are concretized in empirically manifest patterns of social practice at any one time. There is always a potential measure of relative deprivation built into the very nature of experience itself. This is because latent implications can induce new interpretations, new desires, new aspirations. We have also seen that in any society some meanings are sociologically more important than others because they have become institutionalized in various ways as intersubjective presuppositions of practice. And finally we have noted that although all persons interpret their world and negotiate these interpretations, any one person's efforts are limited by a variety of constraints other than those deriving from previously institutionalized interpretations. These other constraints include secrecy, ignorance, competitive variety among interpretations, and mystifications of discourse. One type of constraint to which we have paid close attention is the structure of implications that exist in any given meaning, implications that are matters both of logic and of dominant empirical associations.

The analysis of potential scarcities latent in the structure of existing meanings is one way of engaging in the detection of cultural macrotrends. Trend analysis is not prediction. It is simply narrowing the range of plausible scenarios to sufficiently finite proportions to help us conceptualize large-scale social and cultural "events." The notions of "past," "present," and "future" are themselves only cultural meanings. The determination of "events" and their connections (e.g., the rise of a social movement, the transformation of a regime, the genesis of a war, the decline of an ideology, or the fall of a civilization) is the way in which the scientific observer interprets historical-social time. Events occur at varying rates, and thus many pasts, presents, and futures coexist in the "now." All this means that the human world has a fluidity that makes unpredictable contingencies significant in the lives of persons. Here are needed the skills of the historian whose vocation is to examine the filigree of contingencies in the lives of societies. The sociologist's task is not to predict the future. It is, rather, to provide models of macrocontexts from which microevents derive their significance. Since macrocontexts (e.g., social structures, modes of economic integration, root metaphors, myths, and juristic systems) alter more slowly than do microcontexts, it seems reasonable to speak of long-range trends on the macrolevel with multiple microdirectional possibilities.[16] If one wishes to study trends among these micropossibilities, social research must be designed to address situated actions. It is here that the agenda of phenomenologically oriented social research can make its contribution. It is concerned with clarifying the connecting links between scientific observer models of large-scale sociocultural systems, and the commonsense worlds of participants who live among the alternatives, decisions, and acts of everyday life situations.

The explicit research into such connections, necessary for an adequate nontechnicist approach to social theory, is still sparse. Many investigators do, as a matter of course in their research, assume that persons interpret their world in the interest of discovering its moral intelligibility. And some regard such interpretations as potent factors in historical affairs. However, more than a few also often impose their *own* vicarious interpretations of what the world would look like to *them* if *they* were the people being written about. This move is at times necessary and legitimate. But if done unwittingly (i.e., without some empirical checks or without explicit statement of the assumptions being made), the theories of behavior produced by such researchers will provide both little sense of the creative agency of ordinary persons in world construction and little hope for generating an empirical basis for discriminating rational from irrational agency in human affairs.

Let us elaborate this a bit by illustrating in some detail one important way in which many thinkers have often assumed the world to be interpreted by others. Perhaps because of an occupational bias toward the virtue of intellectual consistency, intellectuals have often assumed that certain meanings are simply incompatible; this assumed incompatibility has sometimes seemed to take on causal status in the explanation of historical change. It is important that thinkers become acutely aware that all theories are themselves instances of the construction of world intelligibility. As such, social theories reflect what the investigator who contributes to them assumes to be morally intelligible, tolerable, or problematic to other people. Like anyone else, he may be wrong.

Do intellectuals have a propensity for interpreting as incompatible things that other people may experience as reconcilable by toleration, compromise, or compartmentalization into morally intelligible order? I have no wish to argue that such a propensity exists to any excessive degree, merely that many expert scholars do not hesitate to project their own (often unexamined) senses of intelligibility into the minds of others. If greater intellectual control could be achieved over this process, several benefits would accrue. First, we stand to gain more empirically accurate knowledge of the spectrum of popular (commonsense) world interpretations and their connections to action. Second, more precise insight would be possible into the *continuities* (as well as discontinuities) between "expert" and "commonsense" interpretations of the world. It is reasonable to hope that these benefits would yield still another: a more precise understanding of the truly *shared* (objective, not "merely subjective") symbolic implications of human environments. It is difficult to see how this could fail, finally, to contribute to the reconciliation of "macro" and "micro" theory perspectives in the social sciences. A social science grounded in such symbolic consciousness is surely less susceptible to technicist seductions than any other.

Some would no doubt say that the intellectual perspective in back of these arguments is philosophical idealism, a Hegelian "cunning of reason." But although ideas have implications, and one cannot haphazardly rename the world at will, there is no singular logic of destiny. If the world sometimes seems too strongly determined, it is because our gaze is fixed on the "present" as a singular consequence of what are really countless interpretive acts of persons, now invisible, whose struggles with their contemporary dilemmas of possibility remain forever unrecorded.

In these last two chapters I have outlined a perspective according to which the sociology of moral intelligibility may be approached from the standpoint of a philosophy of persons. The initial assumption underlying this perspective is that one property of society itself is a persistent strain toward delegitimation of its established routines.

After reviewing legitimacy as a general concept within a phenomenological frame of discourse, the notion of scarcity was appropriated from its more usual materialistic connotations and transformed into the fundamental dynamic through which delegitimation occurs. Scarcities were regarded as growing out of a particular feature of society, the latter defined essentially as a mosaic of meanings. That feature is the capacity of all meanings to suggest to people, by means of implications inherent in them, possibilities for operational concretizations *other* than those actually achieved up to a given point of time. Thus, no actual state of affairs can ever be said intrinsically to satisfy all latent possibilities of meanings for persons who seek to fulfill them through action in history. All meanings are therefore open, pointing in some direction of the "not yet."

Legitimation is an operation of closure. That is, it is an argument for discounting the value of pursuing further implications and for protecting established interpretations by means of enforced social sanctions. Legitimacy is always an unstable artifact of human interpretation—a dike against the never ending trickle, flow, or stream of scarcities. Because no legitimations are such as to be self-evidently applicable to all contingent situations, the interpretation of moral intelligibility is a task that falls to everyone. In that sense, every person is a moral agent, although some groups are always finding ways of concealing this fact from other groups.

The interpretation, negotiation, and legitimation of meanings—intrinsically unavoidable requirements for the maintenance of any human society—are within the capacity only of the human species. This is the basis for the claim that human beings confront a world different in kind from that studied by sciences of the nonhuman. One of the implications of my argument is that sociology and history are really interdependent parts of one discipline. Sociology looks at human action from the standpoint of environments experienced by people as sedimented products of cumulative past actions. These sedimented products (institutions, organizations, intersubjective meanings) are partial determinants of what people at any given time are likely to think, feel, do, or hope for. History, by what still seems common scholarly consent, is not a science of the social. It examines how human agents actually went about interpreting those environments by creating acts and agendas that, in the events that resulted, became part of the life-world of their descendants. Sociologists without historical awareness seem destined to reify the objects of their study, thus obscuring the varieties of human agency. Historians without the sociological perspective often unobtrusively invent it. For the sociological perspective *is* the effort to interpret social reality as something more than a morally haphazard sequence of "happenings."

Notes

1. Macpherson's arguments, some of which I disagree with, are quite complex and cannot be done justice here. A similar approach to the topic is that of Cornford (1972), who identifies Buchanan and Tullock's *The Calculus of Consent* as a major theoretical foundation of a political philosophy of scarcity.

2. To quote Bookchin, "The affinity group could easily be regarded as a new type of extended family, in which kinship ties are replaced by deeply empathetic human relationships—relationships nourished by common revolutionary ideas and practices" (1971:221). Note, in the quotation on page 119, the simplistic confrontation between tribalistic affinity group *Gemeinschaft* ideology and bureaucratic technocracy *Gesellschaft* orientation. Entirely left out of account is the *cultural* problem of tradition and authority. This problem is one of the major foci of the work of Arendt. The connection between her reflections on the cultural organization of society and consciousness, and her concern over the failure of anarchist ideals, is amply seen in her book on revolution (1963). It is this relentless concern for the cultural integrity of nontribalistic authority and Arendt's rescue of the problem from the hands of "grand inquisitor" conservatism that makes her the greatest political philosopher of the twentieth century. It should be added that although anarcho-communism is obviously not the only form of anarchism, it is that form of anarchism that has appropriated the end-of-scarcity theme as its cornerstone.

3. Behind all this discursive variety regarding scarcity stand some very serious issues. The trouble is that many writers reject each other's accounts without necessarily fully appreciating what the main agendas of their interlocutors were. Robbins's discussion of scarcity was part of an effort to undermine what he felt to be an excessively materialistic notion of economics among his colleagues (1946). Polanyi's concern with a substantive economics (versus a formalistic view of economics) was part of his desire to embed the study of economics in a general comparative theory of institutions and social integration. Weisskopff's focus was on the collapse of all relationships between positive economics and normative ethics in contemporary Western culture. The nonmillennial dimension of Bookchin's sociological concerns is the call for a theory of spurious social complexity (1971: 136–137).

4. Aside from its connections with the topic of moral intelligibility there are at least three other compelling reasons why an adequate conceptualization of scarcity is crucial for social thought at this juncture of history. Let us parenthetically note them.

First, we are witnessing the planetwide politicization of the concept of scarcity in connection with modernization and economic development problems. Superficiality of theory in this area can have practical consequences as social scientists increasingly become politically relevant agents of social change. Indeed, scarcity could be thought of as *the* core question in development theories if it is granted that the latter all basically derive from massive changes in people's sense of what they may legitimately expect from life and what they consequently experience as deprivation. (I develop this perspective in Stanley, 1967.)

Second, scientific innovation has made possible the nullification of many heretofore universal constructions of material nature, thereby increasing the range of lifestyle options under conditions of material affluence. It is important to understand that this range is not infinite and in what ways all these options entail a price.

Finally, the concept of scarcity touches on the general problem of linkages between aspects of experience that are accessible to the immediate conscious awareness of average participants in everyday social life and those aspects of life that are not thus accessible. This distinction between observer versus participant models of reality remains an important issue in the philosophy of science. The persistence of scientific modes of explanation that transcend what participants themselves are aware of reflects the conviction of most scientists that *unintended* as well as intended consequences, *latent* as well as manifest functions, and *causes* as well as reasons remain valid concepts in social science. On these grounds, a distinction must be drawn between what I shall call *scarcity* (an observer's claim about some facet of a society or culture) and *deprivation* (statements about some sense of "lack" in the subjective experiences of participant populations in a society).

5. One of the important discussions of this point, in these terms, is Polanyi (1944).

6. This example suggests an unexamined area of exploration for those interested in the metatheory of sociology: the distorting influences of metaphors (obviously, metaphors are not as such distortive). For a highly scholarly, albeit controversial, venture into this domain see Nisbet (1969). For an important treatise in this domain of methodological thought, sure to be influential for a long time to come, see Brown (1977). Specifically on the topic of energy as metaphor in sociology, I owe some deep insights to Marks (1977). Obviously, I also agree with Cornford that "certain goods such as confidence, trust and affection—goods essential to the existence of a political or any other society—are not finite, are not divisible and do not diminish with use, but, on the contrary, grow with their exercise" (1972:37).

7. Robbins, it will be recalled, is the English economist who articulated the now famous "analytical" conception of economics as the "science which studies human behavior as a relationship between ends and means which have alternative uses" (1946:16).

8. So that there is no misunderstanding my argument, I recognize the significant difference between uniformities persisting in time beyond the life span of the individual and the notion of timeless uniformities. My argument requires only the former assumption.

9. For some important discussions of scarcity as a concept see Barnett and Morse (1963), Commons (1961), and Hopkins (1957). A truly beautiful, partly philosophical and partly literary, treatment of the concept of scarcity by an economist is Walsh (1961).

10. Use of "dialectical" is frought with peril, given the chaotic history of the term. I would not place the burden of the word on the reader without an extensive digression explaining what I mean by it were it not for Schneider's helpful essay (and defense of its use) that does exactly this for the term "dialectics" (1971).

11. A note on my use of the terms "implicit" and "implications" is important at this juncture. If the *Oxford English Dictionary* is examined for these terms, it seems that the wide variety of usages it sets forth can be set into two broad classes of meaning: logical and empirical. Statements, concepts, etc., can have empirical implications that one may infer. Thus, given the logical connection that exists

between the concepts of freedom and of choice, any culture that has at its symbolic center the concept of freedom can be expected to "encourage" psychic aspirations that threaten with potential delegitimation all constraints on choice. What sorts of constraints will be under threat at what times depends on the concrete history of the society. Meanings also, however, have a history of material implications rooted in empirical associations between a concept and some one or a few other concepts *or* patterns of action. In this section, for instance, I discuss the implications of the notion of reciprocity as a mode of social integration. Logically implied by the concept of reciprocity is the idea of symmetry. One cannot speak of reciprocity without assuming symmetrical points of reference between which reciprocity is possible. However, I also speak of another implication of reciprocity as a mode of social integration: the notion of "human scale," of community. Community is not a logical implication of the concept of reciprocity. But it is a material implication because of the empirical associations between reciprocal structures of interaction and what people seem to have in mind (and heart) by the term community.

Similarly, consider the often noted connections between Christianity and the history of revolutionary equalitarian movements. Part of these connections is the result of the abstract equalitarianism logically implicit in the Christian notion that all humans are equal in the eyes of God, having been created in His image and endowed with an immortal soul. However, this abstract implication is open to diverse material possibilities of interpreted action. The history of Christendom reflects just about all of these interpretive possibilities, yet there is such a thing as the history of Christendom. A cultural tradition that has at its symbolic center a stress on the eschatological significance of the individual soul organizes the very meanings of consciousness and reality so differently from any other tradition that this tradition, however abstract, becomes one causal influence on how all other aspects of reality are interpreted.

It cannot be my purpose in this chapter to lay out the theoretical assumptions necessary to clarify satisfactorily what causal connections may exist among abstract implications, subjective experience, and social action. This part of my argument must depend on somewhat intuitive understanding of such connections on the part of the reader. But it seems clear that the development of such a psychosociology of culture should be high on the agenda of those interested in the role of symbols in society.

12. This typology requires considerable development. To some extent it has received this in the work of Sahlins (1972:185–314). Gouldner has explored at some length the concept of reciprocity in sociological theory (1973:190–299). And, of course, the literature on exchange theory grows almost by the day.

13. The phenomenological dimension of cultural analysis is not as highly developed as the analysis of social structures. An outstanding example of such an effort is Sahlins's treatment of the philosophical foundations of Mauss's classic essay on gift reciprocity (1972:149–184).

14. As has been widely noted, this dependence and its consequent inefficiencies are probably related to the Soviet Union's occasional efforts to introduce controlled market mechanisms into Russian life. Also see Goodman and Goodman, who provide a proposal for a societal redistribution system designed in the light of modern technology but geared deliberately to avoid the creation of dependency (1960:188–

218). Whether one agrees or disagrees with their particular proposal, this is the sort of performance that distinguishes a fundamental social thinker from the typical technocrat in the field of planning.

15. Eliade stages a beautifully illustrative hypothetical argument between representatives of "archaic" man and "historical" man on the nature of freedom as they see it in the light of their respective phenomenologies of scarcity (1959:154–159).

16. An argument along this line was developed by Hook (1943:246–267). The problem, of course, is how to reconcile culture and action on the level of integral social science theory. See Stokes and Hewitt on the concept of "aligning actions" (1976).

Social Cybernetics

SUBJUGATION BY METAPHOR?

Introduction

Just as generals do not have to rule for a society to become militarized, so cyberneticians need not rule for a society to become technicized.[1] Clues to the technicization of culture can be found in intellectual treatises, in policy vocabulary, and in converging concrete policy practices. Increasing numbers of thinkers are producing a scholarly literature (whatever one may think of its technical quality) promulgating systems analysis as the new "generalist" basis of intellectual inquiry itself. This orientation has its policy literature correlates. A general tendency in much current policy literature is the effort to reduce ends, goals, data, values, decisions, and language itself to instrumental-operational terms. All this is important for the formulation of cybernetically relevant "sociostatic norms."[2] Finally, we find in a wide diversity of specific institutional settings certain tendencies converging toward a common, cybernetically relevant mode of policy and decision vocabulary. These tendencies include the reduction of decision strategies to cost-accounting systems; the transformation of values into quantifiable social indicators; the codification of ends into legislated and measurable accountability criteria; the computerization of centrally stored information banks; and the encapsulation of social life into simulated models of experience.

Such tendencies obscure the differences among the ends, values, experiences, and missions associated with different (and particular) institutions, not to speak of the nuances of life itself. Whatever becomes of these trends, their existence is a contemporary cultural fact. As long as they persist, we must examine the implications of whatever superordinate vision gives them their teleological coherence. That vision is the pantechnicist model of cybernetics. The concept of reality as consisting of cybernetically self-regulating natural systems, adaptive (or calibrated to be adaptive) to their environments, generates a vision of policy design as a potential branch of natural law. In an atmosphere of anxiety and crisis, cybernetics could become the metaphysics of a precarious society driven by the need for total societal regulation.

That cybernetics does occupy this exalted position in the imagination of some thinkers is clear from the sorts of claims that are found scattered throughout systems theory literature. Let us look at five such claims. (1) The cybernetic model subsumes the advantages and traditions of the two master metaphors of Western civilization, mechanism and organism. (2) The cybernetic notion is connected with a major intellectual movement, general systems theory, that constitutes a new intellectual "generalism" capable of organizing and interrelating the various contemporary departments of the intellectual division of labor. (For those who think like this, of course, cybernetics is not a metaphor; it is a theory of reality.) (3) The cybernetic model appears capable of subsuming (rather than displacing) the master legitimating paradigm of liberal societies: the market principle. (4) The cybernetic model is the intellectual foundation of a new humanism that resolves some of the tensions between current definitions of "public" and "private" interests. (5) The cybernetic paradigm is the intellectual rationale for an aspiring social engineering elite who seek to revive, on a non-ideological basis, the nineteenth-century goal of a problem-solving social science.

In light of such claims cybernetics can be seen as either the culmination or the latest stage of trends that have been variously analyzed by some of the foremost thinkers of the twentieth century. Cybernetics may be regarded as part of the history of "rationalization" as analyzed by Max Weber, or of "reification" in Marxist literature, or of the "mathematical conception of things" in Heidegger's treatment of this idea, or of "instrumental reason" as discussed by the Frankfurt school.[3] Even this brief listing of such claims and contexts reveals at least cybernetics' theoretical importance in the Western mind beyond the fluctuations of immediate fashion. What kind of idea, then, is cybernetics? What role does it play in the pessimists' diagnosis of contemporary technicism? Why do some thinkers find cybernetics of such inspirational significance for their philosophical, sociological, policy, and even religious thought? To answer these questions, I shall examine certain intellectual and moral problems inherent in cybernetic systems theory as a symbolization of social order.

I shall proceed in three steps. First, I discuss cybernetics as a general idea. Second, I present an account of the philosophical, sociological, policy, and religious intentions underlying the uses of cybernetic imagery. Finally, reasons are adduced for hesitating to urge the abandonment of cybernetics in social science and policy discourse despite its intellectual fallacies and technicist potential when applied to these areas.

What Is Cybernetics?

One of the difficulties in answering this question is the lack of definitive consensus in the cybernetics literature itself. The usual starting point is Norbert Wiener's notion of cybernetics as the science of communication and control in animate and inanimate systems. But virtually any further reading immediately reveals the problematic nature of such a definition and of its scope. Almost every word in it presupposes controversial concepts and theories. Wiener, and many other figures involved in the cybernetics movement, tend to present their accounts of cybernetics as a history of convergences from disparate starting points toward a common orientation.[4]

Basically we may say that cybernetics is not a discipline but a methodology for the study of communications and control networks in all natural phenomena that appear to an observer to have the properties of self-regulating systems. According to Ludwig von Bertalanffy,

> Cybernetics is a theory of control systems based on communication (transfer of information) between system and environment and within the system, and control (feedback) of the system's function in regard to environment. In biology and other basic sciences the cybernetic model is apt to describe the formal structure of regulatory mechanisms, e.g., by block and flow diagrams. Thus the regulatory structure can be recognized, even when actural mechanisms remain unknown and undescribed, and the system is a "black box" defined only by input and output. For similar reasons the same cybernetic scheme may apply to hydraulic, electric, physiological, etc., systems (1968:21–22).

Bertalanffy treats cybernetics as but one trend in general systems theory along with computerization and simulation, set theory, graph theory, net theory, automata theory, game theory, decision theory, and queueing theory. Other writers see cybernetics as the master concept, assimilating these other trends to cybernetics as the general study of "all possible machines" (Ashby, 1961). Since this has been the main tendency among those who want to adapt cybernetics to the study of human affairs, I shall regard cybernetics in that light in this chapter.

The best way to understand cybernetics for our purposes is to look at it as an intellectual movement with three key metaphysical associations: teleology, process, and holism.

Teleology. Thinkers look to cybernetics to provide a scientific basis for a naturalistic interpretation of purposive behavior relevant to all natural systems, not merely conscious animate (much less human) ones. Few writers deliberately identify themselves with the term teleology. But, as the works of Ervin Laszlo and his colleagues illustrate, the metaphysical thrust of general systems theory (in some places implicit, in others explicit) is against the Cartesian dualist interpretation of mind in favor of a view of mind and purpose as aspects of all "goal directed" natural systems. In the words of L. K. Frank, an early advocate of this ambition,

> We are witnessing today a search for new approaches, for new and more comprehensive concepts and for methods capable of dealing with the large wholes of organisms and personalities. The concept of teleological mechanisms, however it may be expressed in different terms, may be viewed as an attempt to escape from these older mechanistic formulations that now appear inadequate and to provide new and more fruitful conceptions and more effective methodologies for studying self-regulating processes, self-orienting systems and organisms, and self-directing personalities. Thus, the terms *feedback, servo-mechanisms, circular systems,* and *circular processes* may be viewed as different by equivalent expressions of much the same basic conception (quoted in Bertalanffy, 1968:16–17).

Process. Cybernetics is identified with a process view of reality. The master process is "information."[5]

Holism. Cybernetics reflects a commitment to the study of complex wholes in place of the earlier orientation toward "natural phenomena [as a] play of elementary units governed by 'blind' laws of nature." The project is to seek "correspondences in the principles that govern the behavior of entities that are, intrinsically, widely different" (Bertalanffy, 1968:30, 33).

In all discussions of cybernetics a minimum of four concepts appear of crucial significance: "communication," "control," "system," and "goals." The most important point about these concepts is their generality as far as the pioneers of technical cybernetics are concerned. The aim of this generality is to develop methods for dealing with very complex, hierarchically organized systems. Beyond this consensus the precise meanings and domain of cybernetics' constituent concepts become controversial. To demonstrate the lineaments of these controversies would be a project for at least one book in itself. My purpose here is to lay the foundations, with the help of illustrations, for arguing against moves to extend cybernetic concepts into domains wherein their relevance is far less clear than is the case at their points of origin. To do this responsibly, it is desirable to avoid the simplistic, scatter-gun attacks upon technological interpretations of man and society that permeate much of the popular (and scholarly) alarmist literature on technology and values.

There are some abstract criticisms of man-machine models that have been persuasively dismissed to date. For example, it is not fair to set up strict conditions for recognizing "consciousness" in a machine that are still excessively controversial for humans.[6] Second, one should not confuse observations about common properties of men and machines with statements concerning man-machine identity and equality.[7] Furthermore, whatever distinctions exist between men and machines, it is misleading simply to ascribe purposiveness to men and not to machines. According to some notions of purpose, machines could be said to be exemplars of "single-minded" purposiveness compared with men. Finally, and similarly, it is misleading to deny to machines all possible notions of spontaneity, novelty, and creativity. Critics need to be precise about what they wish to mean by these concepts.

More specifically regarding cybernetics, the variation in definitions has been used to argue that cybernetics as a concept is meaningless.[8] This view may fairly be countered by this observation. Those using the term in their technical scientific and engineering work, even when disagreeing about its exact meaning, regard the term as significant. Cybernetic conceptualization has contributed some precise and fruitful models that have illuminated diverse research settings from molecular cell biology, through neurology, to cable engineering. Moreover, it can be shown that some criticisms are not necessarily properly directed against cybernetics as such but may reflect confusions about different uses of cybernetic models. A cybernetic model can be used in at least four different ways.[9] It may be used as a general heuristic device for purposes of precise symbolization capable of generating new research problems and programs. It may be used as a substitute system sufficiently similar in some of its properties to the complex system under investigation to allow for experimentation in place of the original (e.g., a digital or analogue computer showing certain properties of the central nervous system). It may be used for demonstration purposes (e.g., W. Gray Walter's mechanical tortoise designed to demonstrate scanning behavior in a simple mechanism). Finally, a cybernetic model may be used as a prototype blueprint for a system designed for an important practical use. It is possible to criticize a specific model with respect to one or another of these uses without rejecting all of them.

It would be beside the point to conclude from linguistic and conceptual inconsistencies that nothing further need be said. Varied but profound stakes, or intentions, underlie the spread of systems rhetoric. These intentions are of a philosophic, scientific, social, moral, policy, and even religious nature. Reflecting, as they do, the growing desire to unify our speech about all such matters, these intentions will persist despite inconsistencies in efforts to express them in systematic form.

There are moral and intellectual commitments implicit in the language of cybernetic systems analysis.[10] Some of these commitments constitute what I regard to be the legitimate targets of the critics of technicism. Perhaps illuminating them will help the purveyors of systems rhetoric to become more self-conscious about the cultural role they are playing in the symbolic life of contemporary times.

The Uses of Social Cybernetics

SOME PHILOSOPHICAL USES OF CYBERNETIC IMAGERY

There are at least three interrelated projects that seem to lend philosophical and historical significance to the cybernetic movement in science. These projects preceded cybernetics and will doubtless persist even if cybernetics itself declines in its appeal and turns out to be just one more intellectual fashion. Noting these projects should help us understand the appeal of cybernetics to many minds. It will also clarify some of the connections between cybernetics and other intellectual orientations, most specifically, the early twentieth-century energetics movement and contemporary French structuralism.

The first of these three philosophical projects is the search for a unified methodology and a unified language of science around which, despite diversity of disciplines, scientists can cohere as a linguistic and a normative community vis-à-vis nonscientific communities of discourse. The second is the search for a unified (versus a Cartesian dualist) view of the world that can reintegrate man into a broader view of nature without philosophically collapsing the world into mind or mind into matter. What is desired is an "integrated pluralism, an ontology that proclaims both the diversity and the unity of the world" (Bunge, 1969:22). Some thinkers hold out for a type of reintegration that is capable of yielding what is sometimes called a "naturalistic ethic." This means a conception of nature that is somehow capable of ratifying certain values and norms over others, thereby setting limits to the progress of ethical relativism. Third, and related to this second project, is the search for a conceptualization of man in nature that can reconcile warring ideological systems by subsuming them under a common, superordinate imagery.

All three of these sets of aspirations are manifest in cybernetic as well as other systems and structure-oriented literature. Consider the following quotations.

> Energy does not appear as a new objective somewhat, alongside the already known physical contents . . . but it signifies merely an objective correlation according to law, in which all these contents stand. Its real meaning and function consists in the equation it permits us to establish between the diverse groups of processes. . . . As a principle, it signifies nothing but an intellectual point of view, from which all these phenomena can be measured, and thus brought into one system in spite of all sensuous diversity (Cassirer, 1953:192).

> The term social structure has nothing to do with empirical reality but with models which are built up after it. This should help one to clarify the difference between two concepts which are so close to each other that they have often been

confused, namely, those of social structure and of social relations. . . . [S]ocial structure can, by no means, be reduced to the ensemble of the social relations to be described in a given society. Therefore, social structure cannot claim a field of its own among others in the social studies. It is rather a method to be applied to any kind of social studies, similar to the structural analysis current in other disciplines (Lévi-Strauss, 1963:279).

The gain achieved by geometry's development hardly needs to be pointed out. Geometry now acts as a framework on which all terrestrial forms can find their natural place, with the relations between the various forms readily appreciable. With this increased understanding goes a correspondingly increased power of control. Cybernetics is similar in its relation to the actual machine. It takes as its subject-matter the domain of all possible machines, and is only secondarily interested if informed that some of them have not yet been made either by Man or by Nature. What cybernetics offers is the framework on which all individual machines may be ordered, related and understood (Ashby, 1961:2).

These passages from Ernst Cassirer, Claude Lévi-Strauss, and W. Ross Ashby, respectively, span the years between 1910 and 1961. Their contexts are three intellectual movements: energetics,[11] structuralism, and cybernetics. All three movements reflect the projects I have cited. Note that in each quotation the search for a common methodology and a common language takes the form of renouncing the notion that generalization about the content of the world is the end of all inquiry. Rather, it is the structures of relationships (or metastructures) that ultimately matter. Knowledge based upon sensuous relations of human beings to a perceived object world is downgraded in favor of knowledge conceived in terms of conceptual operations performed upon the world. Some antitechnicist thinkers regard this as the ultimate concession, in the very heart of science, to a deeply relativistic ontology and a technological test of truth.[12]

For Cassirer, the paradigm of this shift to the study of relations as such was mathematics: "If we take the mathematical general concept as the starting point and standard of judgment, not only such contents are 'like' which share some intuitive property that can be given for itself, but all structures are 'like' which can be deduced from each other by a fixed conceptual rule" (1953:202). Cassirer himself was rather optimistic about this move because he understood by mathematics the search for a general science of qualities. The significance for him of the "universal energetics" movement was that it carried this intention over to the "totality of physical manifolds."[13] We find the same attitude in Lévi-Strauss, who appropriates it for his brand of structuralism.[14]

Structural studies are, in the social sciences, the indirect outcome of modern developments in mathematics which have given increasing importance to the qualitative point of view in contradistinction to the quantitative point of view of traditional mathematics. It has become possible, therefore, in fields such as mathematical logic, set theory, group theory and topology, to develop a rigorous approach to problems which do not admit of a metrical solution (1963:283).

Cassirer's optimism may be traced to at least two assumptions that he apparently accepted and that are clearly present also in contemporary systems theory. Some critiques of systems analysis as technicist could be said to begin with the rejection of these assumptions. One assumption is that systems theory constitutes a reformed (i.e., a mathematical, non-Aristotelian) formalism. This new formalism challenges the epistemological dominance of mechanistic causality theories. It thus helps combat the Cartesian mind-body dualism and the associated methodological dualism between the natural sciences and the humanities.[15] Cassirer's second assumption is that his move toward formalism resolves the ancient problem of universals and particulars. Here it is instructive to hear his optimism directly. As part of a critique of Heinrich Rickert's pessimism about the cultural role of science (Rickert believed that science's universalizations create an unbridgeable gap between scientific concepts and experienced reality), Cassirer claimed that in the new formalism

> no insuperable gap can arise between the "universal" and the "particular" for the universal itself has no other meaning and purpose than to represent and to render possible the *connection and order of the* particular itself. If we regard the particular as a *serial member* and the universal as a *serial principle,* it is at once clear that the two moments, without going over into each other and in any other way being confused, still refer throughout in their function to each other. It is not evident that any concrete content must lose its particularity and intuitive character as soon as it is placed with other similar contents in various serial connections, and is insofar "conceptually" shaped. Rather, the opposite is the case; the further this shaping proceeds, and the more systems of relations the particular enters into, the more clearly its peculiar character is revealed. Every new standpoint (and the concept is nothing but such a standpoint) permits a new aspect, a new specific property, to become manifest (1953:224; emphasis in original).

We can surely detect in these lines an attitude that appears in our day in Lévi-Strauss's musicology of myth and Ross Ashby's theory of all possible machines. In the intervening years, despite the changing fashions of scientific metaphor, the aspirations I have noted remained stable in all of them. Thus, in Jean Piaget's volume on structuralism (1970), there is the same emphasis upon mathematics as the source of the ideal-typical conception of structure.[16] There is, too, the same optimism regarding possible criticisms of structuralism as a dehumanizing influence. And, finally there is the same hope, once invested in energetics, that structuralism, properly understood, provides the methodological key to a unified, interdisciplinary scientific method and language. In Lévi-Strauss's philosophy of science and method, structural phonetics, synchronic analysis, information theory, kinship theory, and comparative mythology are all pressed into the service of the same quest for a non-Aristotelian formalism that inspired Cassirer.

The polemical battle lines that have been drawn against these aspirations have proven to be similarly stable. In the debate over the relative significance of "semantics" and "syntactics" for the understanding of meaning we can recognize unresolved issues from earlier methodological battles.[17] In arguments such as between Sartre and Lévi-Strauss on history and dialectics, or between Foucault and Piaget on the concept of the human subject,[18] we confront again the age-old question of how properly to assess the place of individual persons in explaining the world and its processes. Finally, in the polemic over the efforts to expand a single idea into a universal scientific method, we reencounter the problem of how to distinguish a fruitful metaphor from distortive sophistry.[19]

Regarding the search for a naturalistic ethic and the reconciliation of ideologies, the terms of reference have changed somewhat over the years. The stakes used to be seen as rooted primarily in the long conflict between religious and scientific accounts of the world. Although, as we shall see later, this debate continues, the conflict between political systems today has taken center stage in the ideological drama. Like structural-functionalism, cybernetics has crossed ideological borders.[20] As we shall see, too, in reviewing cybernetic policy thought, efforts are under way ideologically to reconcile "liberalism" and "Marxism" under a common cybernetic imagery. Perhaps nowhere is this vision of a unifying naturalistic morality so forcefully expressed as near the end of Ervin Laszlo's effort to integrate all thought into systems philosophy.

> The order of nature, as we are now coming to appreciate and recognize it, is the best source of inspiration of social morality we can have. Nothing man-made can compare with it. . . . We are not alone: we are in nature. We are a part of the tremendous, balanced hierarchy which distributes roles and values to all things, from atoms to international federations. . . . No longer is there the great unbridgeable gulf between collectivist Communism and individualist Democracy. The theories clash, but the practical aims are but so many variations on a common theme. The mystic oneness of the Orient, the classless humanism of the Marxists, the group-sensitivity of American youth, are different expressions of a common concern, now emerging into world consciousness: the concern for living together, in systemic wholeness, on a crowded, instantly communicating, and economically and politically interdependent planet (1973:289, 290).

Cybernetic systems theory, then, is associated with controversies that intellectual history shows us are not new. These disputes, which it is helpful to summarize as three interrelated issues, reveal some of the intellectual staging areas for antitechnicist criticism of the cybernetic paradigm. These three issues are the problem of how to reconcile mathematical and non-mathematical senses of concepts like "qualities" and "particulars": the problem of how far to carry the "decenterment" of personal agency in scientific models of the social world and its history: and the problem of how to decide where conceptual "imperialism" begins in defining the limits of scientific models and their empirical domains. We shall encounter these issues in many guises.

SOME SOCIOLOGICAL USES OF CYBERNETIC IMAGERY

As societies become more complex, they seem to generate properties requiring new forms of analysis. Sometimes these new properties appear incompatible with moral standards consistent with earlier forms of society. Sociology, as a modern discipline, can be regarded as part of a struggle to discover the laws or principles that render intelligible the form of society that we now generally call industrial-secular-technological.[21] The major properties of this new form are, of course, large-scale political-territorial organization; complex, technologically rational division of labor; and the rise of the idea of people as statistically analyzable "populations." In democratic societies we also find cultural pluralism, and, modifying the idea of populations, an ideology of utilitarian individualism.

When human beings come to be conceived as "populations" a problem emerges that underlies the program of systems theory. That problem is: how can we understand and control what happens on a large-scale, aggregate level of reference when it no longer seems intelligible to reduce collective phenomena to the sum of individual intentions and actions?[22]

Regarded in this light, the "humanistic" claims of cybernetic systems theory begin to make sense. Note, in the following summary of the intentions of cybernetics, the overarching theme of avoiding an all-out split between macrolevels and microlevels of intelligibility. Put less abstractly, what this means is the desire to leave room, in the explanation of aggregate phenomena, for the intentions and creative acts of individuals and groups.

In its applications to sociology, cybernetic systems theory embodies four intentions:

1. The intention to avoid treating collectivities as classes of ontological entities (organisms, machines, group minds). Instead, cybernetic systems theory is a search strategy for macropatterns of order whose forms are not constrained by the logic of metaphors borrowed from the concrete picturable objects of the empirical world.
2. The intention to show dynamic *interactions* between macrolevels and microlevels of reference. Systems theories stress the importance of contingencies, degrees of freedom, and the emergent nature of macrophenomena.

3. The intention to explain the sources of stability and change within a single intellectual framework.
4. The intention to contribute to the search for an interdisciplinary model of methodology that could eventually heal the rift between "natural" and "humanistic" sciences.

Propenents of cybernetics share the usual scientific antipathy to anthropomorphism. But one of their central intentions is to recast—not reduce—the traditional teleological and humanistic vocabularies into terms that will include the behavior of nonhuman and even nonliving natural systems.

This list paraphrases goals reiterated in many forms throughout sophisticated systems theory literature. It provides us with some means to inquire further into the theoretical nature of the attack that humanistic critics direct against the cybernetic conceptualization of society. If it can be shown that these intentions in some significant way are not, or can not, really be fulfilled, then the critique would become intelligible beyond the level of mere rhetoric.

How may such inquiry be carried out? By the persistent application of two questions: Does cybernetic systems theory do anything that other approaches do not do for sociology? Does cybernetic systems theory obscure anything important that other approaches do not? Let us begin by considering some examples in sociological literature of the uses of cybernetic concepts.

Talcott Parsons has tried to integrate the cybernetic hierarchy of control notion into his structural-functional sociology (1966, 1967). The passages in which these efforts appear, however, suggest more of a flirtation with the terminology than vigorous conceptualization. Parsons says very little that he did not formerly (or for that matter subsequently) say without this terminology. In no sense did he grapple with the implications of the cybernetic apparatus in the ways that he did with other explanatory models like economics or psychoanalysis.

Very different is the work of Karl Deutsch. After an earlier study in which the cybernetic information and communication concepts were tentatively applied to the political sociology of nationalism (1953) Deutsch produced a book, *The Nerves of Government*, whose influence on the ambition of others to apply cybernetics to social science probably cannot be exaggerated (1966). Again and again one finds references to his book as having provided the basic conceptual logic of the cybernetic model as it applies to social science. There are good reasons for this claim. The book is a massive attempt at metatheoretical advocacy for cybernetic concepts. The whole effort is embedded in a detailed review of the functions of models in explanation and in a discussion of some classical models of social thought whose contributions cybernetics allegedly incorporates and supersedes.

Deutsch then goes on to present his arguments for cybernetics by discussing cybernetic operationalizations of a number of concepts like "purpose," "learning," "consciousness," "will and free will," "self," and "mind." An exhilarating sense of optimism pervades the book regarding the "breakthrough" significance for both conceptual clarification and reconciliation between humanistic and scientific concerns. *The Nerves of Government*, however, has the drawbacks of a book dedicated to pure advocacy. There is no reflection in it of opposing points of view on the many specific philosophical problems connected with its major claims. Furthermore, despite the emphasis on the virtues of operationalization, little attention is paid to exactly how these concepts illuminate research problems and findings that have proven difficult to reconcile or even to interpret. Finally, as is true of other would-be pathbreaking books by imaginative thinkers, it is often difficult to tell whether particularly insightful leads have been derived from the conceptual system being defended. Often such leads simply reflect the independent creative imagination of the thinkers without which the conceptual apparatus alone would prove quite shallow.[23]

For example, it is an interesting hypothesis that an egalitarian society might regenerate an unequal status system because "attention and communication overload may force a frantic search for a privileged status for their own messages upon many people." But this image of a democracy as a possible "jungle of frustrated snobs starved for individual attention" could have been as easily derived from a Toquevillian analysis of the logical and cultural implications of equality as from the complex apparatus of cybernetic information theory (Deutsch, 1966:162).

This point applies especially to the extreme emphasis on the feedback concept that Deutsch shares with so many who write in the cybernetic vein. The optimism regarding cybernetic conceptualization generated by the suggestive qualities of the feedback notion can hardly be exaggerated. In Deutsch's volume, and vastly more so in Walter Buckley's synthesis of virtually all process-oriented sociologies under the rubric of systems theory (1967), the concept of "negative feedback" is central. It is the superordinate concept meant to subsume the common properties of organisms and machines, as well as the sociological meanings of stability, deviance, and change.[24] Yet, is the comprehensive character of this concept more than illusory?

After presenting an impressive list of arguments for the claim that the negative feedback concept is superior to a classical mechanistic equilibrium orientation, Deutsch says that "a feedback model of this kind permits us to ask a number of significant questions about the performance of governments that are apt to receive less attention in terms of traditional analysis." What are these questions? "What are the amount and rate of change in the international or domestic situation with which the government can cope?" "What is the lag in the response of a government or party to a new emergency or challenge?" "What is the gain of the response—that is, the speed and size of the reaction of a political system to new data it has accepted?"

"What is the amount of lead, that is, of the capability of a government to predict and to anticipate new problems effectively?" The point of all these questions is that "the overall performance of a political decision-system will depend upon the interplay of all these factors" (1966:189).

We may ask how these questions differ from what would earlier have simply been called political sociology. Perhaps we are quibbling. If cybernetic concepts are useful in stimulating good research, so what? Here let us again ask whether cybernetic conceptualization illuminates anything in a new way or obscures anything that other perspectives do not?

Cybernetic analysis does force our attention, generally, to process questions, to morphological similarities between diverse levels of organization, and to "natural" (unintended) patterns of hierarchy and control.[25] It does so, however, in ways that obscure certain things we should be interested in, things that other perspectives illuminate. The question is: just how necessary are cybernetic concepts for understanding sociological theories or research results? My list of Deutsch's four questions inspired by the feedback notion seems quite compatible with traditional sociological inquiry, and this can be said of his other efforts to link cybernetic conceptualization and inquiry.

But there is more to be said. We are promised that the feedback concept differs significantly from the equilibrium concept as used in structural-functional sociology. Deutsch outlines four differences. First, "in feedback processes, the goal situation sought is outside, not inside, the goal seeking system." But note that we have to cope here with the same problem that we face in the structural-functional conception of equilibrium: the boundary problem. What is the boundary of a human "decision-system" and can this be stipulated precisely enough to determine what we mean by "inside" and "outside"? Deutsch seems to assume, along with many other writers, that there are all sorts of cybernetic systems around. He does not go into the question of who designs the goals and environments in terms of which different cybernetic systems *have* "insides" and "outsides" (Deutsch, 1966: 185 ff.). The problematic nature of this issue is well set forth by John Steinbruner:

> The problem of definitions, essentially that of fixing the scope of analysis, becomes acute here. Cybernetic operations have a distinctly different appearance if one adopts the viewpoint of the designer of the system than they do if one looks at the matter from the viewpoint of the decision-making mechanism within the system. Karl Deutsch, in applying cybernetic models to politics, adopts the viewpoint of the system designer, and in that context he speaks of "goal-seeking feedback" and of a servomechanism which ". . . includes the results of its own information by which it modifies its subsequent behavior." . . .
>
> Such notions are acceptable only in a discussion of overall system and are clearly absurd if the analytic focus is on a particular servomechanism embedded in one of Deutsch's networks. A goal-seeking radar is performing a set of programmed operations which, though drenched in the designer's purpose for oper-

ations in a particular environment, are utterly devoid of purpose as an internal matter. . . . The reason for adopting the lower-order point of view here is that for most interesting applications the overall system cannot be known. That is the meaning of structural uncertainty. The problem is one of understanding how cybernetic decision processes operate when they cannot be set (as by some designer) to produce some desired outcome with complete confidence (1974: 63–64, n. 23).

We shall have repeated occasion to return to this crucial problem.

The second sense in which Deutsch argues that the feedback concept differs from the equilibrium concept is that "the system itself is not isolated from its environment but, on the contrary, depends for its functioning upon a constant stream of information from the environment as well as upon a constant stream of information concerning its own performance." Apart, again, from the problem of system definitions and boundaries, there is the question of what else this generalization means. It seems to mean that the application of cybernetic analysis will increase the rationality of the decisionmaker (whatever *this* be). The British cybernetician F. H. George presents this assumption in deceptively unproblematic form:

> In Management Cybernetics there are two main factors under consideration. The first is very general and refers to the attitude to be adopted, and the second is quite specific and refers to the methods to be used. As far as the first is concerned we have to say that it is in some sense intangible; you have to believe that a system, such as a business or a government, is like an organism and is capable of being controlled by a "brain." The "brain" is the Board of Directors, or the Government itself, and must act in a regulating capacity. What needs to be spelled out is that human brains in individuals and Boards of Directors in companies can be similar. This is not to say that the artificial brain cannot control its system more efficiently than the human brain controls its body; the point is, rather, that if the Board can approach the capacity of the human brain and can also evolve with experience, then it can take full advantage of its ability to make it so (1971:150–151).

Deutsch's discussion of cybernetic rationality is much lengthier than this, but his tone is also optimistic: decision rationality derives from better information processing.

This optimism is perhaps rooted in a somewhat simplistic treatment of the concept of consciousness. Deutsch thinks of consciousness as

> a collection of internal feedbacks of secondary messages. Secondary messages are messages about changes in the state of parts of the system, that is, about primary messages. Primary messages are those that move through the system in consequence of its interaction with the outside world. Any secondary message or combination of messages, however, may, in turn, serve as a primary message, in that a further secondary message may be attached to any combination of primary messages or to other secondary messages of their combinations, up to any level of regress.
>
> In all these cases, secondary messages function as symbols or internal labels

for changes of state within the net itself. They are fed back into it as additional information, and they influence, together with other feedback data, the net's subsequent behavior (1966:96).

Consciousness is thus defined with reference to communication "nets," and in this sense consciousness can be said to apply to all types of nets from a social organization to a nervous system. Deutsch defends the conceptual virtues of the feedback concept of consciousness against its competitors (though not in any detailed, systematic way). The defense, again, turns on the virtues of process analysis. The question of whether consciousness can be interpreted profitably in terms of communication flow, and if so, what exactly this means, is not explored at depth. The persuasiveness of Deutsch's advocacy is based on a combination of these simplistically phrased operational definitions plus caveats in which we are warned about the many complexities of social analysis. The total effect, if one is not careful, is that the sophistication of the caveats seems to lend credibility to the operational definitions as though the caveats were generated by the definitions.

This becomes clear from Steinbruner's intensive analysis of the mutual implications of cybernetic logic and psychological theories. There we see that, contra Deutsch, *from the explanatory and moral standpoint of the rational actor model of decisionmaking*, cybernetic processes appear *irrational*.[26] I shall postpone further discussion of this claim until the section on policy intentions.

The third difference that Deutsch posits between the cybernetic feedback process and equilibrium restoration is that feedback allows for the notion of a changing goal.

> It may change both its position, as a flying bird or an airplane, and even its speed and direction, as a rabbit pursued by a dog. Suitable feedback processes could in principle catch up with a zigzagging rabbit, just as in principle suitable automatic gun directors can track and shoot down an airplane taking evasive action (1966:187).

The fourth difference is that in feedback analysis

> a goal may be approached indirectly by a course, or a number of courses, around a set of obstacles. . . . In a simple form this problem has appeared in the design of automatic torpedoes and guided missiles. In politics it appears as the problem of maintaining a strategic purpose throughout a sequence of changing tactical goals (Deutsch, 1966:187).

The difficulty here is that what looks like conceptual sophistication (recognition that, in the real world, goals change and that there are obstacles) is nullified by a rigidly physical-object notion of what a goal is. (Note, in Deutsch's examples, the metaphor of the chase.) Deutsch is aware of this problem and in subsequent pages he speaks sensitively about the various meanings of goal change. Eventually he recognizes that it is possible to move "beyond the problem of simple goal-changing feedback" to that of "the learning capacity or innovating capacity of [a] society" (1966:199). But it does not seem to me that Deutsch ever finally shows how his elaborations (those of a sophisticated observer of the real political world) can be conceptually reconciled with his cybernetic model.

There is a limited, but for this context perhaps enlightening, parallel between the style of Deutsch's book and Emile Durkheim's *Division of Labor* (1964). Both thinkers regarded themselves as doing science. Both also regarded their work as a contribution to reconciling contemporary scientific principles and traditional moral goals that many other thinkers had come to view as irreconcilable. Durkheim wanted to show that the modern (technologically rational) division of labor was not irreconcilable with the values of individualism. Instead, he argued that the modern division of labor both helped to create and was functionally necessary to the values and norms of individualism. To demonstrate this, Durkheim had to treat as "abnormalities" certain effects of the modern division of labor that others (e.g., Adam Smith, Saint-Simon, and Comte, and after Durkheim, Weber) regarded as necessary correlates of it. Among these were political, moral, and cultural disunity; social class conflict; anomie; and social malcoordination.[27]

In a sense, Deutsch makes the same effort to reconcile classical Western political-moral goals with the latest scientific trend, in this case, cybernetics. As Durkheim spoke of abnormalities in the division of labor that could prevent the emergence of organic solidarity, so Deutsch speaks of "failures or pathologies" that could abort the formation of an "autonomous decision system" (1966:221 ff.).[28] These pathologies are five in number: the loss of system power (resources and facilities), the loss of information intake, the loss of depth of memory, the loss of capacity for partial inner rearrangement, and the loss of capacity for comprehensive or fundamental rearrangement of inner structure.

Despite these potential failures or pathologies, Deutsch's faith remains unshaken in the reconcilability of systems analysis and the efficacy of individual conscious decision-rationality. This faith, as noted, is exactly what is questioned by Steinbruner. Steinbruner's discussion reflects the same austere willingness to face the morally bothersome implications of the cybernetic perspective that characterized Weber's treatment of the theory of bureacracy.

A somewhat more modified optimism pervades the extensive empirical effort to apply the cybernetic perspective to organization theory. Notable here is Jerald Hage's *Communication and Organizational Control*, based on a four-year case study of a community hospital and a comparative investigation of sixteen health and welfare organizations (1974). Perhaps more than any other sociologist, Hage has tried to consider the problems of operationally defining communication as cybernetic control. The entire study is based on the assumption that the volume of communication (and its network structures), quite apart from semantic content, is an important indicator of cybernetic control processes. Hage does not confuse communication with feedback and admits that the existence of true feedback has been assumed and not empirically determined in the study. Hage assumes that successful operationalization of communication as feedback will yield models of functional and dysfunctional structural patterns from the standpoint of organizational survival and adaptation criteria.

Many things could be said about this study, but for our present purposes only the following comments are relevant. The study is not so much about the sociology as the social psychology of cybernetic organization. The bounded bureaucratic organization is the unit of analysis. It is treated as a world in itself. Thus, there is no attention to the societal context of such organizations, their embeddedness in a larger society and culture. Because of Hage's focus and because there is no consideration of the connections between what these organizations "produce" and the environment conceived as a total political economy, the notion of "adaptation" remains undefined. It is the absence of organizational instability (i.e., the presence of steady states) that is the operational criterion here of successful cybernetic control. Hage is completely aware of this, admitting that his study exclusively addresses homeostatic regulation and not the more complex issues of adaptation symbolized in the concept of "moving equilibrium." What is not clear is whether he is aware of just how many issues a purely bounded organization focus permits a sociologist to ignore.

At the end of the book, Hage says

> Cybernetics is clearly comprehensive because it includes all parts of the system, by definition, as well as goals and the environment. The old traditions of structural-functionalism, which can be defined as the interrelationships among structure and performance, goal analysis, technological determinism, and size analysis . . . are all subsumed as different coordinates of problems in a system of organization. . . .
>
> Any system of thought that can combine conflict theory and evolutionary theory in the ideas of instability and moving equilibrium is to be recommended. Revolution and evolution become merely different states of the same system described by different combinations of numbers on the same coordinates (1974: 242, 243).

We see here further evidence of the seductive nature of the cybernetic paradigm. Yet, nowhere does Hage seriously consider the question (and the focus of his study is such that he is not forced to it) of whether there are possible meanings of "communications," "goals," "revolutions," "evolution," and "environment" that are scientifically and/or morally incompatible with their cybernetic senses. What would it mean to consider such a question?

I shall address this issue and summarize my reflections on the sociological uses of the cybernetic paradigm by considering briefly a number of unresolved issues raised by these uses.

SUBSTANCE VERSUS PROCESS. It is clear that the cybernetic paradigm reflects the general shift in contemporary scientific epistemology from substantive to process and relational concepts. In sociology, this shift brings sociological theory into line with a number of broad, interrelated cultural trends: epistemic pluralism (i.e., delegitimation of firm views about the objective nature of social reality); value pluralism (i.e., delegitimation of singular, absolutist moralities); and general ideological relativism (i.e., delegitimation of moral and scientific "authority" vis-à-vis the diverse subjective experiences and orientations of groups and individuals).

Extreme process orientations instantiate in sociology a radical form of the nominalism that has been so prominent a feature of modern liberal social philosophy. As one might expect, this trend has resulted in major debates of mixed scientific and moral nature.[29] None of this means, however, that any consensus has evolved around a single meaning of what the process is that constitutes social life. What most approximates such a consensus is the converging interest in the concept of "communication."

PROCESS VERSUS CYBERNETIC PROCESS. The cybernetic paradigm is really an effort to take the commonsense notion that communication has something basic to do with society and give it firm scientific status and respectability. A successfully applied cybernetic paradigm could do four things for sociological theory. It could consolidate the victory of "process" over "substance" conceptualizations. It could institutionalize the concept of "communication" as the master process. It could give a standard operational significance to the concept of communication that earlier conceptions of communication lack.[30] Finally, cybernetics could counteract the threat of rampant subjectivism that boded a drastic separation between physical science and social science methodologies.[31] Process sociology could have its moral cake (emphasis upon the creative agency of individuals) and eat it scientifically (society viewed as part of the kingdom of "natural systems" all analyzable by way of "information theory").

These are lofty goals, fully relevant to the humanistic stakes set forth at the beginning of this section on sociology. But in one way or another they all rest on adequate operationalization of communication as a cybernetic process. Has this been achieved?

COMMUNICATION VERSUS INFORMATION. In technical cybernetics, "communication" is operationalized as the process of "information" transfer. Most sociologists who write in the cybernetic mode do not seem very aware of the technical commitments involved in the reduction of communication to information. They certainly have not come fully to terms with the fact that "the concept of information in the technical sense provides a purely quantitative measure of communication transactions which abstracts entirely from the interests and meanings of the agents involved" (Sayre, 1967:7). That is, information and semantic meaning are not the same.

In sociology, the importance of nonsemantic, impersonal aspects of

communication is not a new theme. The notion and implications of communication nets have been explored for years in the "structural effects" literature of sociometry. The communications significance of spatial organization has also been a theme of social ecological research. And, as noted
earlier in this chapter, the European debate between the partisans of general
semiology and the philosophers of hermeneutics is relevant to the issues encountered in the domain we are concerned with.

I cannot integrate all these threads here, nor need I. What is clear is that
very few sociological writers in the cybernetic vein have confronted directly
the philosophical problems entailed in operationalizing communication as
information. Those most deeply embedded in this viewpoint write as
though unaware of the challenges that could be brought to bear upon their
use of the cybernetic communication model.[32] A few writers on communication per se do seem aware of certain of these problems, and their writings
contain some effort to elucidate the meanings of communication in order to
determine the proper place, if any, of a cybernetic conceptualization.[33]
Other advocates of cybernetic analysis (especially sociologists) who are
aware of such problems duly present the technical meaning of information
and then proceed largely to ignore the distinctions between information
and communication, between signs and symbols, between syntax and semantics, and the many problems of meaning interpretation.[34]

ARTIFICIAL VERSUS HUMAN INTELLIGENCE.   The artificial intelligence
(AI) movement has directly and indirectly generated a revival of interest in
a variety of traditional metaphysical problems like the nature of language,
teleology, free will and determinism, consciousness, the meanings of randomness, and the ultimate nature of reason itself.[35] Yet here, to a greater
degree perhaps than anywhere else in the cybernetic literature, one finds
people thinking past each other. To some extent this is unavoidable; the
range of issues to be addressed, from the philosophy of action, through the
philosophy of science, to the philosophy of language (not to speak of the
sciences proper) is simply immense.

But much of the confusion is clearly the result of the contemporary nature of education itself, its separation into airtight domains of specialization
with no superordinate domains of discourse stressing a common heritage of
philosophical problems. Having reviewed a reasonable amount of relevant
literature on all sides, one thing seems evident to me. The advocates of the
cybernetic paradigm in sociology are far from making a foundationally persuasive case. Most especially this is clear as regards the problems of conceptual equivalence generated by the AI movement (purposes and goals in
human as against natural systems, communication as against information,
free will as against randomness, action as against behavior, groups and persons as against systems and nets, and the various meanings of consciousness, language, and interpretation in humans as against machines). These
problems have been all but ignored in the social science applications of cy-

bernetic analysis. To the extent they are, the charge of technicism, at least by default, is surely justified.

PERSONS VERSUS INDIVIDUALS. I have left for last the most significant unresolved problem as regards the pessimistic interpretation of cybernetic systems analysis: the theoretical status of the human.

One of the humanistic claims of systems analysis based on process theories, as we have seen, is that the input of individuals is taken account of, allegedly more so than in earlier macrosociological theories based on organic, historicist, or structural-functional perspectives. At the same time, some theorists in the systems mode either directly or indirectly celebrate the removal from scientific center-stage of the subjective, conscious intentions of persons. This downgrading of the explanatory significance of the conscious self and its intentions is clearly present in Lévi-Strauss's adaptation of cybernetic information theory in anthropology. In a different intellectual context this downgrading is present to an almost cultic degree in Foucault's semiological view of history (1975). In still another context the same dismissal of the complexities of the human self is present among many AI formalizers: "What do judges know that we cannot tell a computer?" asks John McCarthy, a leader of the AI movement (quoted in Weizenbaum, 1976:207).

Commenting on the problem of the individual's theoretical status in structuralism, Piaget has this to say:

> It might seem that the foregoing account makes the subject disappear to leave only the "impersonal and general," but this is to forget that on the plane of knowledge (as perhaps on that of moral and aesthetic values) the subject's activity calls for a continual "de-centering" within which he cannot become free from his spontaneous intellectual egocentricity. This "de-centering" makes the subject enter upon, not so much an already available and therefore external universality, as an uninterrupted process of coordinating and setting in reciprocal relations. It is the latter process which is the true "generator" of structures. The subject exists because, to put it very briefly, the being of structures consists in their coming to be, that is, their being "under construction" (1970:139–140).

In a different context, Piaget assures us that

> even perception calls for a subject who is more than just the theatre on whose stage various plays independent of him and regulated in advance by physical laws of automatic equilibration are performed: the subject performs in, sometimes even composes, these plays; as they unfold, he adjusts them by acting as an equilibrating agent compensating for external disturbances; he is constantly involved in self-regulating processes (1970:59).

The trouble is that the reassurance is too general relative to what the humanist critics of technicist theorizing have in mind. Reassurances such as Piaget's do not address the technicist challenge to a moral philosophy of persons inherent in cybernetic systems theories. The point is not just that the human individual participates in the construction of social systems. The question is: *how* does the individual participate? Does he participate authentically? inadvertently? knowingly? intentionally? autonomously? in a mystified state? functionally?

Just to list the code words for the ideological disputes that have raged over how to define the moral person is to reveal the unresolved status of these disputes. To raise the question of participation regarding cybernetic systems theory is to begin to move out of the strictly scientific intentions of cybernetic systems theory into the area of policy intentions. For if one is puzzled by the continued attractiveness of the cybernetic paradigm despite its intellectual ambiguities, part of the answer lies in what it seems to promise with regard to managerial policy interests.

SOME POLICY USES OF CYBERNETIC IMAGERY

To understand the place of cybernetics in policy analysis we must recall the distinction between parochial technicism and pantechnicism. This distinction is helpful for appreciating the differences among three somewhat distinct categories of policy literature: "systems analysis," "policy science," and "managerial cybernetics."

Parochial technicism is a state of mind in which the world is viewed myopically through the lens of one particular technique (and its interests, agents, and organizations). Pantechnicism occurs when the entire symbolic world is reconstituted as one interlocking, problem-solving system according to the latest technologically relevant language of control (at present cybernetics).

Parochial technicism is now very widespread in Western industrial civilization. It is a natural outgrowth of the centrifugal effects of a technological division of labor that divides society into specialized compartments of perception, experience, and action.[36] Likewise to be expected in a technological civilization are countervailing efforts to create the intellectual foundations for an integral, comprehensive pantechnicism. The planning-engineering dimension of macroeconomic theory can be viewed as a step in this direction. So can French technocratic positivism from Condorcet to Comte, which sought to use the emerging science of sociology to create a social technology of controlled progress.

Parochial technicism is a partially pluralistic phenomenon. Its common features are a faith in techniques of some sort, a confidence in technical expertise as such, and an assumption that all important problems are technological. But otherwise many instrumental flowers bloom in the technological garden. Pantechnicism has an important additional feature that contradicts even the minimally pluralistic nature of parochial technicism. Pantechnicism, as the term suggests, is grounded in a singular (the pessimists say intellectually totalitarian) interpretation of the world. At minimum it is a claim that a universally effective problem-solving methodology exists (e.g., cybernetics systems analysis and planning). At maximum, pantechnicism is grounded in an ontological claim (e.g., the world *is* a vast cybernetic system). There is a certain tension possible between the parochial, somewhat pluralistic sort of technicism reflecting competing grounds for expertise and the generalist singular vision underlying a pantechnicist view of the world.

The modern history of policy theory and practice subtly reflects the tension between these two kinds of technicism. We can see it reflected in the rhetoric of "systems analysis," "policy science," and "managerial cybernetics." Examples abound. American defense planning provides a fruitful range of them.

In the latter years of World War II the air force began to plan for its postwar future.[37] The orientation underlying this planning was a model of parochial technicism.

> The men involved did not consider the identification of post-war enemies, the efficacy of an international organization, and the foreign policy goals of the United States irrelevant; but they considered these factors of secondary importance. First priority throughout the entire planning period went to the creation of an independent air force, second to none in size, technological advancement, and strategic capability. If this goal could be accomplished, the planners were sure that no matter what political situation the United States might face it would have the necessary military forces to carry out its policies (Smith, 1970:107).

This degree of technicist parochialism was unmistakably of the presystems analysis variety.

"Systems analysis" did not appear in American defense planning until the advent of the RAND Corporation and its influence. In the form that it did emerge it was an extension of the operations research methods developed in World War II, rather than what has since come to be known in academic circles as general systems theory. So in this less recent sense,[38]

> any discussion of systems analysis logically begins with the fact that economics increasingly has come to play a major role in solving many major defense problems. To the systems analyst, decisions about alternative weapons systems, force structures, and strategies are essentially problems of economic choice (Sanders, 1973:13).

"Policy science" is a catchall term for certain efforts to create a separate policy analysis-technology-planning discipline.[39] Policy science is commonly thought to include policy-relevant social science findings plus techniques like systematic alternative future conjectures, trend projections, simulation, and social indicator construction. The difficulty with this literature is that its programmatic qualities have far outstripped its conceptual and empirical dimensions. There is a disingenuousness in the American policy science literature. This disingenuousness arises from the literature's extreme moral abstractness. All sorts of desirable goals are abstractly set forth, caveats about the difficulties of achieving these goals are presented, and a successful policy science is depicted as a discipline that can reconcile the goals and the caveats. There is very little conceptual analysis of the possibilities for such reconciliation and equally little consideration of the few empirical studies that bear critically upon the project of building such a discipline.[40] The literature assumes, as a sort of prime directive; that policy science must (and therefore can) reconcile democratic processes with a scientifically oriented expertise of societal guidance and management.

Few American policy science advocates are outright Saint-Simonist in the technocratic sense. Indeed, what gives policy science its surface plausibility, as contrasted with the style of systems analysis cited earlier, is the rhetorical attention to the importance of political factors in policy analysis and management. Policy science texts are so full of warnings and qualifications that it would be difficult to apply to their authors what is sometimes said of systems analysis: "Current usage suggests that he who has 'systems capability' can analyze, engineer and manage any system." (Hoos, 1972:18).

There is a striking change in all this when we move to the explicitly cybernetic oriented policy literature. Those who identify themselves as policy-relevant cyberneticians are advocates of a truly comprehensive pantechnicism. The main point that policy cyberneticians make over and over again is that cybernetic management is an intelligence amplifier. What does this mean?

The systems analyst, thinking from within the rational actor model (presupposed in analytic decision theories), seeks to provide tools to increase the individual actor's means-ends rationality. The policy scientist wants to produce analysts who, in addition to being "bright, non-conventional men of high analytical capacities . . . should exhibit maturity, explicit and tacit knowledge of political and administrative reality, imagination and idealistic realism as well" (Sanders, 1973:293; this is a paraphrasing of Dror's views). Both of these approaches address the problem of making the rational actor more rational. Both seek to provide actors with tools for more intelligent individual decisionmaking (tools like cost-benefit analytic techniques, gaming, alternative future projection, and simulation).[41]

However, cyberneticians begin by assuming the individual actor's insurmountable limitations of intelligence relative to existing problems of social coordination and control. In managerial cybernetics this is uniformly the starting point of policy analysis.[42]

> People have mistakenly assumed that modern problems of control in the social, economic and industrial spheres are within human capacity to solve—or can be brought within it by a process of education and learning. This is the fallacy with which cybernetics is ultimately concerned (Beer: 1964:130).

Stafford Beer, perhaps more than any other writer in this vein, tries to abort his reader's temptation to reify analogies between man and machine into notions of total simulation and "thinking machines." The intentions are laudable. Yet when Beer describes cybernetics as "a concept of control far in advance of anything hitherto considered by managers and engineers" and as "a new vision of society" that emulates the key features of biological systems, he tacitly contradicts his own caveats (ibid.:207).

In his 1973 Massey Lectures, Beer defended cybernetics against charges of technocratic bias by presenting a cybernetic conception of worker participation, rather than rule by engineers, accountants, or operational research men.

> The people who know what the flows are really like are the people who work in the middle of them: the work people themselves. And if their interest can be captured in putting together the total model of how the firm really works, we shall have some genuine worker participation to replace a lot of talk about worker participation (1974:41).

But this is not the point. Technicist attitudes and concepts are vividly present in Beer's conception of social modeling. These include viewing society as patterns of technologically manipulable communications flow; the downgrading of even enlightened personal judgment; and the analogizing of "control" processes in human social organizations and biological organisms.

> Then contemplate a company that is run from a control centre, in which the dynamic flow chart, continuously reflecting the world outside by teleprocessing data into it, constantly holds the pattern, and uses the computer constantly to monitor all that variety. We are near to this concept in running a battle, or a warship, or an electricity supply system.
>
> The vision I am trying to create for you is of an economy that works like our own bodies. There are nerves extending from the governmental brain throughout the country, accepting information continuously. So this is what is called a real-time control system. This is what I mean by using computers as variety handlers on the right side of the equation. They have to accept all manner of input, and attenuate its variety automatically. What they will pass on to the control room is whatever *matters* (ibid.:42, 43; emphasis in original).

This is all, of course, incredibly apolitical, ahistorical, and asociological. There is no hint in such rhetoric of the real world nature of political, class, or other interest conflict, no insight into the political nature of "information," no respect for "information's" ambiguities of meaning deriving from moral paradoxes and situational diversities.[43]

Beer is not unaware of possible criticism. But in the name of the cause his awareness remains markedly unsophisticated, the defense often resting on specious distinctions.[44] His argument that cybernetics is compatible with individual freedom is uninformed by any visible sense of the fundamental cultural, constitutional, and conceptual revolutions that such a reconciliation would require in industrial democracies (such as those implied in my earlier consideration of a libertarian technicist model). The radical down-grading in cybernetics of the humanistic hope for the political significance of "seasoned men of good judgment"[45] is likewise ignored by Beer and other managerial cyberneticians. This is because, for societal cybernetics as for corporate cybernetics, the overall management problem is the same: "It is for the company to go on living. The subsidiary problems have to be solved in the context of survival" (Beer, 1964:133). The totalitarianism implicit in philosophically unexamined technicist models often rises to the surface suddenly in a passage, only to submerge again in pious hopes.

> And obviously, the question arises, whether an autonomous firm [or an individual person] will agree to collaborate in such a scheme. The reply is that government has many inducements to offer in obtaining the information it needs, and the greatest of these inducements is the fact that industry cannot expect sympathetic treatment from government policy if it will not contribute useful and timely information (Beer, 1974:42).

We must turn to the Soviet cybernetic literature for a less obscured confrontation with the totalitarian implications of cybernetic technicism.

It is ironic indeed that the most vigorous ideological resistance to the claims of cybernetics, resistance justified by unresolved philosophic problems about conceptual equivalences, reductionism, and the downgrading of historical dialectics, should have occurred in the West's most totalitarian society. Ironic but not paradoxical, because a totalitarian society generates an official ideology whose stability must be secured in the face of all potential competitors. Cybernetics was perceived as one such competitor. It was not until 1954 that attacks on cybernetics ceased in popular Soviet publications. By 1961 cybernetics was sufficiently acceptable that at the twenty-second congress of the Communist party "the need was voiced for expanding cybernetic studies in order to create the material-technical base for Communism."[46]

The flirtation with the image of total societal control that one finds in the cybernetic literature of liberal democratic societies emerges without coyness in the Soviet version. It is, of course, true that the ideology of voluntarism is still sufficiently important in the culture of the liberal West to set rhetorical limits to visions of perfect regulation. The shading varies depending on whether one speaks from within what looks like a very individualistic society or one that is self-evidently authoritarian. Thus, Amitai Etzioni mutes his reader's "hyperactive" emphasis on voluntarism by reminding us that

> while the language of societal action draws on voluntaristic concepts, it rejects the hyper-active postulate of voluntarism. We view constraints on societal action as neither tentative nor abnormal, but as an integral part of our meta-theoretical perspective. These constraints arise in part from the fact that each macro-action unit is composed of a multitude of micro-units whose relations and ac-

tions affect macro-action; we also assume that, as a rule, there is more than one societal actor—i.e., that there is more than one macro-unit which is endowed with the capacity to act and, hence, also with the capacity to limit actions of other units. In addition, the actors are usually related in ways not subject to their control. In this sense, a collectivistic element is added to voluntarism in constructing a language of social action (1968:71).

Meanwhile, V. G. Afanasyev, one of Russia's foremost academic cyberneticians, feels obliged to make sure that his readers understand that

> at the same time, each member of socialist society also exercises control. He is a co-owner of public property, genuine master of the economy and society as a whole, and so acts the part of manager of the entire economic, social and political life. In the last analysis, man under socialism is also in control of any fractional system, an enterprise, let us say, and takes part, in some degree or another, in managing the affairs of the community to which he belongs (1971: 79).

But the Soviets do not obscure the directive, centralist Saint-Simonist element of cybernetic thought. Although they speak of a socialist society as a "self-controlling" one, they do not mean by this simply "the control processes objectively inherent in society." Toward such processes their position is unequivocal: *"The free market with its anarchy and competition is replaced by scientific planned control"* (Afanasyev, 1971:83. His italics.) And what is meant by scientific control?

> Scientific control . . . of socialist society means man's conscious purposeful action on the social system as a whole, or on certain of its components, based on knowledge and utilization of its objective laws and tendencies with a view to providing optimal conditions for its functioning and development and ensuring that it should achieve its goal. . . . It means to ensure the integrity of the system, assimilate or neutralize internal and external disturbance and continually improve the structural and functional unity of the system (ibid.:83–84).

Soviet disingenuousness does not take the form, as in liberal cybernetics, of making invisible, through abstraction, the sources of social control. No hairs are split about the authority of the Communist party replacing the market as the foundation of "the conscious, purposeful action of the system of government and other agencies." Soviet disingenuousness is to be found, rather, in the mystification of totalitarian repression connected with the history of Communist authority. There is no trace of any sense of irony, tragedy, or evil, nothing but moral self-righteousness, in assertions like these:

> Another significant feature of socialist society as a self-controlling system is that control under socialism is directed at creating a new communist society, the material and technical basis of communism, at forming communist social relationships, educating man as a versatile personality.
>   Under socialism, however, the aims of individuals coincide in the main with the ultimate result, therefore the latter does not loom over them as a blind hostile force but comes as an anticipated, programmed result, an aim achieved. The

social, political and ideological unity of the members of socialist society makes it possible to design a common goal and muster effort to achieve it. As a result, the laws of socialism are manifested not through the actions of isolated individuals, as happens under capitalism, for instance, not through antagonistic class struggle but through the joint action of all members of society united by feelings of friendly cooperation (Afanasyev, 1971:86, 88).

Interesting, too, is Afanasyev's conception of "objective conditions" and "subjective factors" in cybernetics. In the liberal democracies social cyberneticians must contend with a political ideology of "participatory democracy." This ideology stresses the political significance of every citizen and the educative significance of all institutions for the development of personal competence.[47] Leninism, with its emphasis on the vanguard role of the Communist party in all things, aborted any such bourgeois individualist trends in Soviet communism. With the replacement of traditional historical materialism by a cybernetic theory of development and control, to judge from Afanasyev's work, we may expect increased emphasis upon technicist authority and a concomitantly severe attack upon "mere" subjectivism.

Any system has internal and external objective conditions which the subjective factor has to reckon with. By itself, however, the subjective factor does not coincide with the conscious activities of all those belonging to one system or another but is the expression of the activities of those who fulfill control functions. In other words, with respect to control the subjective factor is the activity of the subject of control in providing for the optimal functioning and development of the system, it is the activities of those who are competent and authorized to make decisions.

Subjectivism is a very dangerous thing. It leads to hollow arbitrary decisions that cannot possibly be fulfilled. It entails an overstrain on material, human and financial resources, causes disorganization and undermines confidence in the government. In effect, subjectivisim interferes with the normal, let alone, optimal, functioning of society, breeding disproportions, giving rise to a tendency to make out that things are better than they are, and so on (1971:91, 198–199).

The totalitarianism that remains implicit in such a vision of control in the liberal democratic literature is made quite explicit in this text:

The Communist Party works out a single political line to be pursued in every sphere of the country's life and conducts the organizational, ideological work for its realization in practice.

Indeed, the CPSU as a governing party is called upon to effect guidance in all spheres of social life, the spiritual sphere among them.

Developing socialist democracy certainly does not mean encouraging idle talk and anarchy, insubordination and general forbearance. One must also remember that development of democracy is inseparable from the development of the socialist state and socialist legality. Nobody should be allowed to violate Soviet laws under the flag of democracy or use the prestige of public opinion to cover up anti-social acts. (Afanasyev, 1971:125, 129, 249).

We find here, too, as we do in the liberal cybernetic literature, a bias against complex and ambiguous communication (i.e., against natural language).

> It is an indispensable condition of optimal control that the communications should be rational, simple and easily controllable, that the controlled and controlling systems should have no unnecessary components, nor superfluous intermediate links in their communications. Cybernetic devices, computers and automatic devices among them, are a powerful means of regulation, of "reduction of entropy" in social systems.
>
> Lack of authentic information is a source of subjectivisim, of purely volitional and arbitrary decisions and actions incompatible with scientific control. Hence the collection and processing of information and its best possible utilization are an indispensable attribute of control (Afanasyev, 1971:175, 182).[48]

Cybernetic self-regulation also supplies a rationale for the Marxist notion of the withering away of the state. But it is evident that the price, to liberal democratic ears, is the replacement of the "particular wills" of class and state by a most singular "general will":

> Highly developed production and community life under communism cannot exist without strict order, without a single will directing the efforts of millions towards a single goal. Production, work will always be the mainstay of social progress, and social labor will always have to managed. Under communism the role of planning, organization and regulation, and especially of accounting and inspection, will be greater than ever before. (Afanasyev, 1971:251).

And what justifies this totalitarian reduction of the lives of persons to computer-digestible information useful for the "application of mathematical methods to the cognition and administration of society"? Here Afanasyev, following Lenin, rises to true theological heights. Recognizing that "mathematics by itself, without any special sociological studies, is unable to lay bare the qualitative nature of social phenomena and so become a potent instrument of scientific control," he calls upon Gödel's incompleteness theorem. According to this theorem, "one cannot completely describe a probabilistic system using no other evidence than that derived from the system itself." But—no problem in a socialist society. The need for transcendental meaning is solved: "Mathematics finds this necessary evidence in Marxist sociology, which reveals the nature of social phenomena. Only after this is done, do figures acquire a real meaning" (Afanasyev, 1971:258, 259). Quantity and quality are reconciled in Marxism. The ironic circle is complete. "Socialist" and "bourgeois" managerial cybernetics are sanctified alike in the transcendent realm of theology. For to some in the Christian democratic West, too, cybernetics represents "a system of redemptive incarnation" (Desmonde, 1971:60). And so we move, inevitably it seems, to the possibilities of cybernetics as religious revelation.

SOME THEOLOGICAL USES OF CYBERNETIC IMAGERY

In the long history of Western religious literature there are many ways in which ideas about man, gods, and nature intertwine. I shall examine two of these. One reflects the intention to reconcile the vocabulary of natural science to that of theology. Without stretching such labels in too procrustean a manner, I shall borrow for this the term "Thomism" after Thomas Aquinas, whose great achievement was the reconciliation of contemporary natural philosophy (Aristotelianism) to medieval Roman Catholic theology. The other genre of literature I shall explore seeks to relate man, gods, and nature by doing the opposite of the first: reconciling theological concepts to science. Let us borrow for this the term "naturalism." In the first, then, the theological world view is ontologically superordinate to the scientific, and in the second the relationship is reversed.

This little classification is useful because it illuminates some of the religious intentions behind the adoption of cybernetics as a theological vocabulary. Theology is an indicator of the religious imagination's yearning to conceptualize the sacred organization of the world. It provides a clue to the unresolved spiritual tensions of an age. Seen in this light, cybernetic theology, no matter how intellectually problematic, has important things to tell us about the status of optimism and pessimism in our time.

CYBERNETICS AS THEOLOGICAL THOMISM. Cybernetic Thomism is a small but intellectually significant movement in contemporary theology (the more so if one agrees with those theologians who would assimilate Teilhard de Chardin to this orientation). At its most ambitious, cybernetic theology seeks to provide a comprehensive account of the realtionships between mechanism and revelation, the predictable and the unpredictable, in terms of the contemporary apparatus of systems logic. This ambition is notably evident in passages such as the following:

> On a broader, cosmic scale, the [cybernetic] model might be considered as part of a process whereby God redeems the natural world. Starting with some type of undifferentiated substance or energy, God iteratively informs this primal matter. In accordance with archetypes (algorithms) within the Divine Mind, the original substance acquires forms. Individual entities arise out of the void as successively more complex *Gestalten*. This individuation process is an iterative incarnation of the Logos, occuring through a dialogue between God (the non-determined part of the system), and the natural world (the determined part of the system). Each discrete entity arising in this process is an algorithm embodied in a material form. The entire universe is the process of enrichment in the Mass. Man assists in the reforming of nature by imitating God (Desmonde, 1971:61).

The essay from which this particular passage is drawn is an effort to use Gödel's theorem and the theory of Turing machines to argue that "the universe must be a mixed mechanistic–non-mechanistic system." A mixed system is one in which "the mechanistic part of the system utilizes information to transform itself into a more comprehensive algorithm." This "iterative enrichment process" is regarded as the scientific meaning of what salvationistic theologies stress as "reversion to a source." Thus, "The iterative reversion process includes biological evolution. The species man assists in the redemptive process by attaining visions of the relation of the One to the Many. These visions are expressed in axiom systems and in attempts to construct universal calculi." The implications of this for the resanctification of technology are immediately evident: "Technology originates in icons (imitations of divine exemplars) with which man meditates on the relation of the One to the Many. Technology is basically a type of alchemy in which matter is transformed into a higher form." And thus, through cybernetic theology, "the development of science and technology is placed within the context of the redemption of the natural world" (Desmonde, 1971:73, 74).[49]

The generic program of cybernetic Thomism is perhaps nowhere better expressed than in this passage by a theological thinker whose project quite literally is the reconciliation of cybernetics and Thomistic theology:

> St. Thomas Aquinas did not disdain to use the models and explanations of the physical science of his time. Certainly the contemporary Thomist cannot decently ignore contemporary science. The philosopher as a member of the human community must strive for relevance. What is more relevant to the contemporary world than the effort to penetrate the limits of operationalism in the sciences, and to reopen to modern men the intellectual legitimacy of a model of the universe that may conceivably include spiritual realities not open to empirical verification instrumentally conceived (Dechert, 1965:5)?

The optimism that typifies this mode of theology is present in Dechert's seemingly uncritical faith that, in addition to unifying physical science, social science, and public policy, cybernetics also promises to modify traditional Marxism and to "provide a bridge between the Christian humanistic tradition and the contemporary scientific-operational mind."

Interestingly, Dechert stresses the possibility of cybernetics fulfilling the lapsed program of the energetics movement.

> Even the revolution in physics at the turn of this century succeeded only in dematerializing the universe to the extent of converting "solid" matter to energic relations. As brilliant a mind as Henry Adams could equate, as basic frame of intellectual reference, twentieth-century "energy" with thirteenth-century "form" and only conclude that the universe as conceived by men has moved from unity to multiplicity. Indeed it had—not because energic relations do not exist in reality in all their multiplicity, but because scientific thought had not reached a point where its conceptual tools could re-integrate the relatively unformed primordial particles and energy to which it had analytically reduced the universe (1965:11).

Cybernetics enables us to recognize and articulate the

> Anti-entropic process by which multiplicity is ordered and patterned into higher

unities—in which process a reduction in available energy may be accompanied by increased complexity of structure and by an increasing span of autonomous functional activity (ibid.).

Thus, cybernetics gives us nothing less than a perspective capable of restoring to the "scientific mind a conception of hierarchy and a principle of order"(ibid.:14)[50]

The optimism here is so prevalent that it overshadows caveats that might lead in other directions. So it is that Dechert comments, almost in passing, that

> a critical problem confronting the student of information is the development of modes of analysis whereby not only is a patterned configuration rendered intelligible and its information content quantified, but whereby judgments may be rendered objectively, and perhaps quantitatively, *on the relative ontological content and value of such formal configurations.* [Aquinas, Dechert reminds us, said that] " . . . the order of things demands that all things be divinely arranged in a proportionate way. God made all things in weight, number and measure" (1965:16).

One important reason for the optimism in analyses like Dechert's may be worth noting for its more general significance. Despite the title of his essay, Dechert's analysis of the person is distinctly nonsociological, remaining primarily concerned with the conceptual elements of a cybernetic and Thomistic psychophysics of consciousness.[51] If the problem of reconciling a cybernetic and a theological philosophy of persons is conceived on this nonsocial and virtually nonmoral level, then a degree of optimistic exuberance makes sense. Much of the pessimism regarding cybernetics is influenced by historically (even eschatologically) oriented social philosophy and a highly action-oriented moral philosophy of persons. In Dechert's discussion of information, consciousness, and soul, he points out that in the view he is examining regarding the reconcilability of these concepts, "Descartes' *cogito* . . . is far more than a circular argument giving a ground for his philosophizing. It is an ultimate statement of the human personality and the ground of individual human identity" (ibid.:29). A sociologically grounded theory of identity, of course, rejects this position in favor of a philosophy of socialized action ("I *interact*, therefore I am.") From this angle, approaches like Dechert's do not even begin to address the moral meanings of personhood in a cybernetic sociology.

CYBERNETICS AS THEOLOGICAL NATURALISM.   Thomism seeks to reconcile scientific and theological truth, with the latter predominant. Naturalism represents a shift of emphasis in which whatever is regarded as the scientific attitude toward truth becomes the dominant criterion of validity. As natural philosophy became ever more "naturalistic," ending in science as we know it now, religion was accounted for in increasingly naturalistic terms, culminating finally in a purely functional interpretation. According to this interpretation, the realm of the "supernatural" is a set of illusions functional to states of the "natural" world. Cybernetic naturalism reveals itself in efforts to assimilate supernatural (theological) categories to natural (cybernetic) categories.

For perhaps the most obvious example of this orientation let us return to Deutsch's *Nerves of Government* (1966:228–244). Deutsch develops a sort of survival theology of natural autonomous systems. His efforts to redefine theological categories in cybernetically relevant terms are inspired by a question that forms the chapter subheading introducing his theological reflections; "How can autonomy be protected against failure?" He treats such perennially interesting topics as humility and pride; lukewarmness and faith; reverence and idolatry; love, cosmopolitanism and nationalism; grace and curiosity; and eclecticism and spirit. There is neither time nor need to review all Deutsch's efforts in this vein. One example, Deutsch's treatment of the concept of grace, will illustrate what is going on.

Deutsch introduces the notion of grace by first discussing curiosity as a human or animal system's orientation toward outside information and conditions. There are two consequences of curiosity that can result in failure of system autonomy, and it may be for this reason that there has been "an element of reserve in the treatment of curiosity on the part of many ethical or religious thinkers." One consequence is system "drifting" that can occur when information overload interferes with system steering capacities. The other harmful consequence of curiosity is the treatment of outside data "as if they had no claim on us at all." This situation occurs when the relevance of information as beneficial or hostile to system autonomy is not attended to because the information is treated with "cold curiosity or the esthetic thrill of the mere spectator." That is, "Whatever we discover or experience is then treated as a mere instrumentality, a source of pleasure or excitement, or a mere tool for the accomplishment of goals it can neither set nor change" (1966:236).

In contrast, the concept of grace stands for an attitude toward "information or events originating from outside ourselves as answers to our innermost problems of self-determination." Such events are not treated as mere instruments, relevant only to narrow subsystem goals, nor as blind forces that drag us helplessly along.

> Rather they are available to us as improbable events or data that may offer us the missing pieces to our puzzles, the particular crucial elements needed to resolve a particular inner crisis of our decision system (Deutsch, 1966:236).

This all requires a certain state of preparedness, an openness to new information, and the capacity constantly to assess its significance for the steering autonomy of the whole decision system. In complex systems, this attitude is directed also within, toward new "combinational possibilities and ways in which . . . remembered data could be recombined into new patterns."

All this adds little to the theology of grace, but it subtracts a great deal. Although Deutsch claims that his treatment transcends narrow instrumentalism, it does not, as any comparison with phenomenological accounts of faith and grace experiences (such as Kierkegaard's, to take but one example) immediately shows. Decision system autonomy is not the same thing as salvation. Curiosity disciplined by a sense of system relevance is not equal to the struggle for faith in some source of hope, a struggle fought daily by the religious imagination on a battlefield of doubt and contrary evidence. Finally, Deutsch's whole discussion strains the reins of metaphor. No amount of talk about systems can obscure the fateful difference between such talk and the religious person's existential solitude in the face of finitude, morality, and doubt. Curiosity, faith, hope, or grace are personal dramas, whether of victory or of defeat. They have reference to the spiritual domain not to the organic. It is through human personal experience that we come to use such terms; all else is metaphor. Deutsch's cybernetic vocabulary would have it the other way around. It cannot be so. Religion, whatever its functions, is ultimately not about system survival but about human dignity. Although dignity has to do with a form of autonomy, not all forms of autonomy have to do with dignity. The grounding of religion in an abstract rhetoric of system autonomy completely obscures distinctions between "human experience" and "system behavior" and hence between human dignity and system survival. That such confusion is possible in our day is a profound source of pessimism among the critics of contemporary culture.

A more sophisticated cybernetic naturalism is that of the American theologian Herbert W. Richardson (1967). Richardson embeds his presentation in two significant methodological assumptions. One is that there is no such thing as permanent public atheism, only temporary forms that are signs of the obsolescence of one symbolic representation of the divine and the transition to another. With reference to the present situation, he asserts that "modern atheism is not the culmination, or perfection, of the modern world, but a sign that the modern period of history is coming to an end." The other assumption is that "the beginning [not the final stage] of all true criticism is the criticism of religion, that is, of 'holy ultimates' and 'sacred myths'. And because public atheism is this kind of criticism of religion, it is also the beginning of all other criticism" (ibid.:5).[52]

Richardson believes that "an atheistic culture is impossible in principle." This is because "every intellectus requires a religious foundation if it is to sustain itself as the principle of a cultural epoch" (1969:29).[53] Richardson does not clarify exactly what he means by "requires" (e.g., is the god the representation of the symbolic order of the intellectus itself, or is it Plato's noble lie, Comte's church of humanity, Durkheim's mythical transformation of society itself, or the authority of Dostoevski's Grand Inquistor?). Tempting though it is to engage Richardson's whole provocative, broadscale effort to create a distinctively American theology, we must confine ourselves to considering his place in cybernetic thought alone.

For Richardson, the modern intellectus is organized around the idea of the person, and it is this intellectus that is doomed by new collective conditions of society. Hence, current public atheism is directed against the deity appropriate to this intellectus.

> The significance of modern atheism is not in its denial of conceptions of God that are appropriate to other cultural epochs, but in its denial of the God who is appropriate to our own cultural epoch: that is, the God who establishes the infinite value of individual personality, the God who undergirds democracy and capitalism, the God who makes empirical science possible (1967:10).

Richardson extensively acknowledges the unique institutions and perspectives generated by the modern intellectus. One of these perspectives is the "person" versus " nature" division in which

> both the certainty of moral judgments and the nonmoral character of "nature" are protected by establishing morality in some experience or activity of personal self-consciousness. Hence, modern morality is never *natural*, but is always *personal*—for nature is the realm of mere "facts" and values are now rooted in the person alone (ibid.:12; emphasis in original).

But now, in addition to our personal self-consciousness and our natural science knowledge of the world, a third sort of knowledge is beginning to dominate the intellectual life of our age. This new kind of knowledge is one of control, the object of the sociotechnical disciplines.

> By sociotechnics is meant the new knowledge whereby man exercises technical control not only over nature but also over all the specific institutions that make up society: i.e., economics, education, science and politics. Hence, sociotechnics presages the end of "economic man" as well as of "scientific man" and "political man." It replaces these separated institutional functions with the cybernetic integration of society within a single rational system (Richardson, 1967:16).

Richardson attacks the twentieth-century critics of pantechnicism by asserting their irrelevance. He accuses the critics of ignoring history and not acknowledging the "inevitability of the transformation that is taking place." The argument of inevitability is based on Richardson's assumption that the old arts and sciences are helpless before the critical problems of our age, "problems the modern world has generated but has not been able to solve." It is the need to control these problems that has given rise to the sociotechnical disciplines. These disciplines constitute "a more profound and penetrating rationality" which is overthrowing the modern intellectus.

Richardson's theological program is wholly adaptive to this allegedly inevitable development.

1. Theology must develop a conception of God which can undergird the primary realities of the cybernetic world, viz., systems. And ethics must reorient its work in terms of these systems and focus on the problem of control. . . .

2. The total cybernetic system must be fortified by an eschatological sym-

bolism which can provide it with general goals and assist men to make the continual transitions an increasingly complex system requires. . . .

3. A sociotechnic theology must develop new ethical principles which will enable men to live in harmony with the new impersonal mechanisms of mass society. This ethic will affirm the values of a technical social organization of life in the same way that earlier Protestantism affirmed the values of radical individualism and capitalism. . . .

4. Theology must create new liturgical forms and new myths whereby the unity of sociotechnical life can be presented and experientially "felt" by all men . . . (1967:23, 24, 25–26).

Richardson's theological technicism is based upon an inevitability thesis so absolute as to preclude any critical role for theology in a pantechnicist civilization. This is true to the extent that Richardson is apparently able to accept with a straight face "professional sports" and "mythic heroes of comic books and television as Batman and Superman" as important "new forms of liturgy which will support and clarify the meaning of sociotechnic life, to give man a "feel" for its meaning" (ibid.:26). Against those who might worry about the quality of such a civilization, Richardson disassociates himself as a theologian from critical responsibility for any of it: "my proposal that Christianity affirm and shape a sociotechnical intellectus is not based on any preference for this intellectus, but on a recognition of its inevitability" (ibid.:29).

There is, of course, no warrant in contemporary science for such a deterministic account of social development. Even if there were, it would constitute only an empirical generalization. Theology supposedly deals with the normative life. Overlooking the difference reveals again the status of cybernetic naturalism as a theology of social survival. Ironically, there is a tantalizing passage, left completely undeveloped, in which Richardson seems to be ready to warn us about something.

Not only can Christian theology undertake this task, but it must undertake it. The sociotechnical intellectus requires the same religious foundation that the individualistic intellectus required in order to be redeemed from its own peculiar demonic tendency (1967:29).

What that tendency might be is never stated (neither the terms "evil" nor "demonic" appears in the book's index), nor is it clear whether "redeemed" might include any form of liberation from the agencies of pantechnicist social control.

Perhaps Richardson is trapped in an anachronism that, as a Christian, he does not wish to face: the possibility that the modern intellectus is so deeply rooted in Greek-Hebrew-Christian dimensions of Western history that the end of modernism is also the end of Christianity.[54] This possibility has occurred to other theologians who have turned to Oriental traditions for the religious implications of what they sense in contemporary systems theory.

This ecological view of man as an organism/environment is as foreign to the Christian view of the embodied soul as it is to the popular materialistic view of man as an intelligent biological fluke in a mindless and mechanical world. Ecology must take the view that where the organism is intelligent, the environment is also intelligent, because the two evolve in complexity together and make up a single unified field of behavior (Watts, 1964:9).

In the view of the late Alan Watts, who was one of the most dedicated theological exponents of Hindu and Buddhist religious thought in this country, modern scientific thought is describing in three main ways "the identity of things or events as the mystic feels them."

> The first is the growing recognition that causally connected or related events are not separate events, but aspects of a single event. . . .
> The second is the tendency to think of the behavior of things and objects as the behavior of fields—spatial, gravitational, magnetic or social (ibid.:226, 227).

The third scientific orientation Watts regarded as congenial to the mystical view of things is Ludwig von Bertalanffy's general system theory. Watts devoted eighteen books to articulating why he considered contemporary intellectual trends in the West point beyond theology, beyond atheism and nihilism, and also beyond the Western sanctification of personality. Can it be that the ultimate destination of cybernetic theology is the Oriental mystical vision of unity?

> A superior religion goes beyond theology. It turns toward the center; it investigates and feels out the inmost depths of man himself, since it is here that we are in most intimate contact, or rather, in identity with existence itself. Dependence on theological ideas and symbols is replaced by directed, non-conceptual touch with a level of being which is simultaneously one's own and the being of all others. For at the point where I am most myself I am most beyond myself. . . . It is simply an assertion of the perennial intuition of the mystics everywhere in the world that man has not dropped into being from nowhere, but that his feeling of "I" is a dim and distorted sensation of That which eternally IS. . . . I have been trying to suggest all along that this is what one must come to by following the Christian way intently and consistently until one realizes the full absurdity of its (and one's own) basic assumptions about personal identity and responsibility. (1964:224, 225).

Is a Nontechnicist Use of Cybernetics Possible?

Given the problems I have identified with the uses of cybernetics, it would be tempting to close this chapter by simply rejecting cybernetics as a bad idea except in the most technically proper domains of its application. Yet, at least as metaphor, cybernetics has captured the imagination of serious thinkers. It symbolically organizes certain ways of looking at the world that are in line with much experience in modern society and culture. The cybernetic model does seem to constitute a challenge to the theoretical, moral, and policy imagination. Is this challenge significant? Is it merely "fence-sitting" not to urge right here the abandonment of cybernetic imagery in the areas of its application that were explored in this chapter?

Such might well be the conclusion of those who dismiss systems philosophy as *nothing but* an ideology, functional to the many-faceted interests of technocrats and their patrons. This view seems reflected in the following comment, for example.

> As metaphysics, systems philosophy advances modern philosophy not at all, remaining mired within the Cartesian dualism of mind-body, with endless reiterations of the world view of positivism; it might, in fact, be described as the last agony of positivism, except that its protagonists see themselves as triumphant everywhere. . . .
>
> If systems theory is not a coherent theory, nor usable technique (except in artificial, limited situations), then what is it? It is, of course, an ideology, and one easily understood by considering the changes that have occurred in American (and world) society since the 1940s. . . .
>
> Systems theory, then, can be understood as an ideology—a claim to power and other social benefits for a specific class called by Galbraith the "technostructure"—in which the claim is disguised in terms of the public good, what Marxists might call a new mystification offered by the new technocrats grasping for power (Lilienfeld, 1975:657–658, 659).

I do not regard this position as adequate.[55] It does not address many of the supraideological intentions discussed in this chapter. Despite its vulgar applications, cybernetic systems theory is a potentially important tool of the human imagination. What its strictly scientific value will turn out to be awaits far more sophisticated work than has yet been done. To judge that here would be premature. But in a somewhat looser sense, cybernetic imagery has significant practical and scientific value in its functions as metaphor. It focuses our attention as nothing else quite as effectively does on the need to study scientifically the real-world correlates and determinants of three values that are of great importance to modern secular humanity. These values are "complexity," "spontaneity," and "survival." It is the interaction of these phenomena both as values and as facts that generates the preoccupation with control in modern societies and with cybernetics as the scientific study of control phenomena. Let us briefly see what this generalization means in a bit more detail.

If modern men were exclusively concerned with social control, they would sacrifice complexity in its name. The persisting communal movements reflect the willingness of some persons to do just that. Yet complexity facilitates the technologically rational appropriation of nature for human ends. Complexity also allows for social and cultural pluralism within a single, large political order, for personal anonymity in dense, heterogeneous population centers, and for the pursuit of varied work goals made possible by the liberation of most people from subsistence labor. In short, complexity has various connections with human freedom.

Yet complexity, as a social fact, also generates some unanticipated structural properties that have consequences experienced as threatening to the very values that complexity has also helped concretize. As a general science of complexity, cybernetics studies those aspects of society that are to be understood as consequences of complexity. As such, it should be regarded as a limited but possibly indispensable instrument of humanity's control over its own incarnation as social organization.

As everyone knows, spontaneity in the abstract sense is highly valued in modern culture. To the untutored imagination, spontaneity is the very opposite of control. But, of course, it has long been understood that such is not the case. Classical economics is founded on an interest in the relationships between spontaneous action and the impersonal, self-regulating, control mechanisms generated by it (the system here conceptualized as the "market"). The antitechnicist wing of classical economic theory represented in the literature most ably by Friedrich Hayek and Ludwig von Mises is informed by an abiding faith in the "invisible hand"of the free market as providing the most morally optimal balance possible between spontaneity and control in the affairs of men. The most systematic assault upon that faith, theoretical Marxism, reflects a different interpretation, but the same awareness, of the intimate relations between spontaneity and control in society. Marxism focuses upon the process of hierarchy formation and solidification arising from human productive intercourse with nature. Their well-known differences stand forth so sharply precisely because Marxists and classical liberals are united in their concern with the benefits of spontaneity as a dimension of human freedom.

It is not surprising, then, that cybernetics should inspire hope for a more refined science of the balance between spontaneity (action as the stuff of process) and regulation (stabilized feedback loops). It is difficult not to believe that in some form cybernetics could contribute to our better understanding of this connection. At stake, as systems theorists constantly remind us, is the preservation of the idea of *impersonal* controls (the handmaiden of spontaneity) in place of personal controls (trial and error authoritarianism). In this light, any attack upon the technicist authoritarian manifestations of cybernetics could be considered a possible contribution to its limited appropriation as an instrument of freedom.

Finally, as everyone by now at least abstractly knows, intercourse between humanity and nature has evolved to a point where the planet we inhabit is reacting as a system to human technological actions upon it. Human survival may depend upon understanding planetary processes as cybernetic systems. Only such knowledge can help us integrate ourselves into our planet's natural order rather than exhaust its resources. The newspapers are daily filled with examples of this general issue. Cybernetic imagery knows no boundaries; political, social, or ideological. It focuses our attention on the systemic parameters of life on the planet as a natural order.

In this way cybernetics forces us to reflect upon our identity as a species. Properly utilized, cybernetics could aid us in preserving our human as well as our species identity by showing us how to optimize the flexibility of the natural order we inhabit through standards of planned stewardship. After all, in an age of planetwide technology, this is the only operational way to speak of freedom and survival simultaneously.

Given these possibilities, is our concern for the technicist misuse of cybernetics perhaps overdone? Are the examples in this chapter mere eccentricities of enthusiasm on the part of people just discovering the virtues of a new metaphor? I believe not. Reasons for my caution range throughout this book. They are integrated by a sense of the long, slow, and fragile evolutionary history of freedom and its concretization on the level of the person. In the context of this chapter, it may be helpful to defend my sense of caution by anticipating two major objections that could be raised to my critique of cybernetics: first, by contrasting cybernetically oriented policy analysis with the concept of personal agency, it might be thought that I am implying conscious authoritarianism on the part of would-be social engineers oblivious to the pluralistic nature of modern democratic societies; second, stress on the concept of personal agency could be interpreted as equating the dignity of personal agency with an argument for the equal validity of all *opinions* found in a given population. (Such an equation would reduce the concept of dignity to hypocrisy since all serious people know that opinions are not equally valid and that it would be dangerous to treat them as though they were.) Let us address these objections one at a time.

THE FIRST OBJECTION.   It is possible to argue that cybernetically oriented social engineering rhetoric has totalitarian potentialities without denying that its exemplars recognize the existence and abstract legitimacy of social pluralism. This is because there are ways of handling pluralism that are consistent with a totalitarian approach to decisionmaking.

One way to facilitate totalitarian decisionmaking is to pay lip service to pluralism while encouraging the steady reduction of meanings to the status of quantifiable operational indicators for the sake of creating an easily manageable and machine-analyzable *public* criterion of meaning. Money is the paradigm of this process. To the extent that such indicators replace other criteria of meaning, such a solution can be called *decision by reduction.*

A second approach to decisionmaking that can reconcile the rhetoric of pluralism with the drift to cybernetic technicism is to redefine the principles that legitimize constraints on pluralistic variety. There is, after all, no such thing as a fully or purely pluralistic society. There are always constraints. Though difficult to define exactly, these constraints are rooted in certain broadly shared but abstract principles such as definitions of reality, norms of propriety, the strategic importance of certain institutions, and some conception of national purpose. At a time such as today, when even these shared meanings break down and when pluralism appears to some to border on anarchy, people begin to worry about what the new constraints should be that could arrest the pluralistic drift before it does indeed lapse into anarchy. Such anxiety could facilitate an uncritical acceptance by the public of new constraints. Among these new constraints could be the somewhat arbitrary interests and values of social engineers—these interests and values disguised by abstract talk of system equilibrium, disturbance, feedback, and cybernetic controls.[56] Since the public is unprepared to evaluate the credibility of such technical language, this tendency could be called *decision by technocratic mystification.*

Finally, there is a third approach to pluralism that is consistent with a totalitarian, technicist drift. Pluralism, as has been noted, is something of an abstract term. It does not carry with it its own referent, the unit of analysis referred to. Thus, one can point to pluralism in social practices, social interests, values and norms, and styles of life. Perhaps the most firmly established unit of analysis in liberal democratic culture is the pluralism of *opinions.* Whatever other disagreements there are, almost everyone would insist on the right of people to "their own" opinions.

There are two dangers in grounding pluralism in the markets of opinion. Consider the distinction between "opinion" and "judgment." I regard it as useful to restrict the term opinion to the expression of untutored logic, compartmentalized levels of abstraction, extreme subjectiveness, and "ready-at-hand" justifications. Judgment is then definable as an orientation resulting from thoughtful application of certain procedural rules to experience. These rules are designed to validate an opinion according to principles considered justifiable by the judger not by reference merely to his "own" opinions but by higher order notions he believes to be legitimate and rational in some more lasting historical sense. In short, people appear more casuistically careful with their judgments than with their opinions. Untutored opinion is therefore more irrational and hence more easily manipulated. It is, moreover, a potent force when it is politically managed because of its intimate association with emotionally charged symbols. To define pluralism *merely* as the right of untutored opinion to exist, therefore, is to invite subversion of personal agency by state managers through selective utilization of information. This may be called *decision by propaganda.*

THE SECOND OBJECTION. This review of three approaches to decision-making that seem to reconcile the rhetoric of pluralism with the totalitarian potentialities of social engineering enable me to address a second objection to the arguments of this chapter. This, it will be recalled, is that my emphasis upon personal agency could be equated with the sentimental but easily disprovable notion of the equal validity of all opinions. It is not opinion that is threatened by cybernetic uniformity so much as the capacity for and the relevance of personal judgment.

As the distinction between opinion and judgment I have just drawn implies, there are more and less valid, more and less tutored approaches to human values, emotions, and norms, however difficult such determinations might be in particular cases. The deeper, more lasting relativities of experience are not rooted in the infinity of mere opinions but in the more finite range of variation in fundamental value systems of the sort that underlie cohesive "life-styles." Something as profound as a "style" of valuing and "being-in-the-world" has a certain internal logic or cohesion of its own. This has been taken for granted by serious thinkers through the ages. In our times, thinkers as diverse as Sören Kierkegaard and Max Weber wrote extensively on this point; Weber in his work on the sociology of religion, and Kierkegaard in his delineation of the inner contours of the aesthetic, ethical, and religious ways of life.

As criterion for action, then, intensity of opinion is entirely unrelated to wisdom of judgment. Wisdom is best pursued by informed meditation on the history of the human race. To be informed, in this regard, means to know something about and have wrestled with the meanings not only of events but also of myths, religions, arts, sciences, and biographies. It is this heritage of clues into the qualitative possibilities of the human experience that should provide the standards of judgment to be applied to all particular ambitions to mathematicize discourse about human intentions. I am not arguing against applied science in appropriate circumstances. But I am warning about its use as a substitute for the education of human judgment. To consider it a substitute is to abandon the historical development of the free person as an evolutionary project.

These remarks should clarify why I avoid categorically rejecting cybernetics, simulation, or any other possibly useful way of analyzing collective phenomena. The point is to maintain the status of analytic models as tools, as instruments for the illumination of personal existence. It was, after all, Norbert Wiener himself, who said:

> For the man . . . to throw the problem of his responsibility on the machine, whether it can learn or not, is to cast his responsibility to the winds, and to find it coming back seated on the whirlwind. . . .
>
> I have spoken of machines, but not only of machines having brains of brass and thews of iron. When human atoms are knit into an organization in which they are used, not in their full right as responsible human beings, but as cogs and levers and rods, it matters little that their raw material is flesh and blood. What is used as an element in a machine is in fact an element in the machine (1954:185).

Systems theories can contribute to the explanation, and perhaps the efficient operation, of actual societies. From the standpoint of Western humanism, however, the ultimate *moral* test must be the degree to which such theories provide tools for the refinement of insight into rational personal agency and its constraints. The ultimate danger of systems theories is that they might provide the most seductive of all opportunities yet conceived for what Sartre has called "bad faith" (1966:86–116). Bad faith means the projection of responsibility unto socially created symbols (be they roles, nations, or systems); falsely concrete objects whose agents *we* then become.

In B. F. Skinner's famous novel, *Walden Two*, considered by many a major technicist document, the narrator (representing Skinner) asks his friend Castle, a philosopher, if he thinks Frazier—the founder and social engineer of Walden Two—is a fascist. (1948). In reply Castle, although stating that he does not know the answer, criticizes Frazier for overlooking all sorts of questions in his social engineering doctrines: "What about the dignity and integrity of the individual? Where does that come in? What about democracy? . . . And what about personal freedom? And responsibility." Skinner has his narrator comment to himself on this, "So far as I was concerned, questions of that sort were valuable mainly because they kept the metaphysicians out of more important fields." (ibid.:242).

Pantechnicism, as science and ideology, is a vision of natural systems struggling for survival. To the extent the human drama is about more than this, then, contra Skinner, it is time again to speak of dignity.

Notes

1. The title phrase "social cybernetics" is important to keep in mind when evaluating the arguments in this chapter. I do not want to be read as guilty of presuming to attack cybernetic systems theory in its physical science and engineering applications. In these domains, the effects of cybernetics have been illuminating and useful. My concern is exclusively with its *uncritical* importation into the social sciences. (As I try to show at the end of the chapter, I do not oppose its critical uses for limited analytical purposes.)

2. This term is Schick's and he means by it norms that are "the servomechanistic trigger of public action" (1970). This concept differs from the usual sociological conception of norms. Schick's idea is that there is a class of norms whose major significance is that they trigger system functions. The model is that of a self-regulating, servomechanistic system. Presupposed in this model applied to planned human societies is that the system has been designed with certain goal states in mind so that some norms are taken as standardized indicators of movement toward or away from these goals. Naturally, compared with a true servomechanistic system, terms like "trigger" are clearly metaphorical. The closest approximation of this approach to norms is the classical economic market model, whose "laws" can be though of as talking about norm parameters that "trigger" system reactions. Schick points out, "A cybernetic state operates under sociostatic norms: employment rates, poverty levels, educational criteria." For norms to function in this manner they must be translated into the quantitative-operational terms appropriate to the cybernetic no-

tion of meaning as "information." Thus, it is clearly appropriatie to regard the "social indicators" movement as a stage in the evolution of a cybernetic society. Schick's essay is one of the very few efforts known to me that seeks to distinguish bureaucratic from cybernetic society. Its focus is primarily the legal sociology of this distinction.

3. Max Weber used the concept of rationalization in diverse ways. For an overview of his work as integrated by this concept see Bendix (1960). For Heidegger's view, see Heidegger (1967). For the concept of "reification" see Lukács (1967). For the Frankfurt critical philosophy orientation see Habermas (1971).

4. See Wiener (1965) and McCulloch (1974).

5. Bertalanffy (1968:22). For a sophisticated treatment of technical information theory accessible to the educated lay reader see Massey (1968). One of the best general treatments of the conceptual terminologies relevant to cybernetics is Sayre (1976). The technical meaning of "information" has to do with the reduction of uncertainty in a mathematical, nonpsychological sense. As Sayre says:

> Although information may signify such different matters as notification, knowledge, or simply data, in any case the imparting of information is the reduction of uncertainty. . . . In this more specific terminology, we may define information simply as increased probability. This definition makes possible the direct application of probability theory in the development of other concepts with related technical meanings (ibid.:23).

The technical description of the information content of an event is the number of times its initial probability must be doubled to equal unity (i.e., actual occurrence).

6. These examples are drawn from Hatt (1968:171–175).

7. For this problem see the excellent discussion by Beer (1964: pt. 4), where he discusses the cybernetic analogues of fabric, mechanism, uncertainty, and language.

8. The next few comments are indebted to Apter (1966:chap. 1).

9. This classification is from ibid.:10–20.

10. I use the term "cybernetic systems analysis", but the reader should be warned that semantic diversities exist in the use of these terms. For a review of some of these diversities see ibid.:chap. 1. For differences between cybernetic theory and general systems theory see Bertalanffy (1968:3–29, 149–151). Since most who have tried to apply systems theory to social science have not distinguished to any great extent between cybernetics and systems theory, I shall use the term cybernetic systems theory whenever it appears reasonable to assume that those whose work I am analyzing intend these terms as synonyms.

11. The energetics movement presents an interesting chapter in the comparative history of metaphoric transfer in science. See Elkana (1968) and Hiebert (1966). Just as Lévi-Strauss did with information theory, White tried to harness the energy concept to general anthropological theory (1959). For the role of the energy concept in American sociology earlier in this century see Eubank (1932: chap. 10).

12. See the account of these changes and Lévi-Strauss's defense with regard to structuralism by Gramont (1970).

13. Cassirer was aware of challenges to the position he took on mathematics as a language (1953:202–203, n. 69). Although heightened in sophistication, the debate over the status of mathematics as a language in social science and social policy is

hardly new to the twentieth century. See Baker (1975). For an important critique of the position from which Cassirer was operating see Dreyfus (1972).

14. On this question of the relations between the concepts of quantity and quality as applied to the role of mathematics in the social sciences see the various essays in the special issue of *Daedelus* (1959). See also the essays by Lazarsfeld and Spengler in Woolf (1961).

15. It is not surprising, therefore, that Cassirer understood and accepted the implications of Bertalanffy's move in 1932 to systems analysis in biology (Cassirer, 1960:168 ff.). For Cassirer's optimistic reasoning about the overcoming of Cartesian dualism and the overcoming by modern (not Aristotelian) formalism of the methodological split between science and humanities see ibid.:162 ff. For the influence of the desire to reintegrate the world by way of a comprehensive and scientifically respectable philosophy upon the personal intellectual development of a leading contemporary systems philosopher see the preface to Laszlo (1973). Laszlo's whole book represents his effort to concretize this desire.

16. Of course, in the intervening years mathematical development led to refinements in the notion of formalism that enabled Piaget to distinguish between "structures" as self-regulating systems and other sorts of forms, a distinction obviously relevant to cybernetic theory.

17. For critical but fair discussions of the limits and ambitions of the semiology movement in general and Lévi-Strauss's place in it in particular see Broekman (1974); Pettit (1975); and Jameson (1972). For a sensitive account of the connections between the baffling diversities of human experience and the desire for a universal human science in the genesis of Lévi-Strauss's thought see Geertz (1973). See also Jameson's distinction between a philosophy of the sign and a philosophy of the symbol (1971: 223–225).

18. See Lévi-Strauss (1966: chap. 9) for Lévi-Strauss's own views of his controversy with Sartre. For exposition of the controversy with relevant bibliography see Rosen (1971). See also Abel (1970). For a helpful entreé to Foucault's thought see White (1973). For the *Esprit* interview with Foucault in 1968 that presents, in a nutshell, Foucault's somewhat overwrought but fascinating conceptualization of the de-centerment of the human subject see Foucault (1975).

19. For a discussion of this see, for example, Runciman (1973).

20. For the migration of structural-functionalism to the Soviet Union see Gouldner (1970). For an example by an Eastern European sociologist of an effort to formalize the syncretism of functionalism and Marxism see Sztompka (1974).

21. This is evident in 100 years of sociological classifications stressing premodern as against modern times. These classifications are so heavily dichotomous as to suggest a persisting consciousness among sociologists of all history as a division between contemporary (industrial-secular-technological) times and all that came before.

22. There is a sense in which this is always a problem, even on a level of three-person groups. But the problem is, of course, tremendously complicated in a mass society that holds as a prime moral value the political input of each individual. As has been clearly understood from Plato to modern utopians, the social macro-micro

distinction becomes a virtually insurmountable dualism (both intellectually and practically) when the community of citizens exceeds a certain size. This topic has in recent years become an explicit focus of scientific attention in what is called the collective decision literature.

23. Readers usually are willing to grant an imaginative writer defending a comprehensive point of view credit according to his own account of the source of his insights.

24. The feedback concept is used to analyze stability in Hage (1974); to analyze deviance in Scheff (1966); and to analyze change in Maruyama (1971). For a listing of the conceptual virtues of feedback see Deutsch (1966: chap. 11). For the technical history of the feedback concept see Mayr (1970).

25. For example, see Pattee (1975) and Whyte, Wilson, and Wilson (1969).

26. Although Allison (1971) never mentions cybernetics, his Model II explanation for the Cuban missile crisis fits closely Steinbruner's (1974) assimilation of cybernetic theory and cognitive psychology. In both cases (Allison's missile crisis analysis and Steinbruner's NATO Multi-Lateral Force treaty decision situation), this cybernetic approach seems to illuminate the irrational nature of decisions when contrasted with the implications of the rational actor model, be the actor an individual person or a government.

27. Durkheim, the reader will recall, subsequently changed his mind about this strategy. Hence his lifelong concern with the scientific, moral, and policy problems of anomie.

28. Deutsch also makes an interesting effort to assimilate religious terminology to his cybernetic model (1966). I review this effort in the subsection on religious intentions.

29. These debates have affected not only abstract issues of sociological conceptualization but concrete policy implications as well. Most notably this is true of the general concept of deviance and its specifics (mental illness, crime, suicide, etc.). See Matza (1969); J. Douglas (1970a); and Scott and Douglas (1972).

30. Consider the great variety of meanings of communication underlying the theories of thinkers like George Herbert Mead, Harold Innes, Marshall McCluhan, Paul Ricoeur, and Hugh Dalziel Duncan.

31. According to Schur,

The sharp distinction between social and natural sciences is implicit in . . . the emphasis on subjective meanings as a key element in social action; in the concern with actors' commonsense knowledge of everyday life; in the expressed preference for methods geared to *verstehen*, or general understanding; and in the view of social science as an order of secondary constructs. . . . We may recognize the importance of "intersubjective understanding" in human social action, but this recognition may well constitute more of a stumbling block to the social scientist than a guideline for his analysis. Although Schutz and others have argued vigorously that recognizing the unique features of each social situation and of a particular individual's subjective experience of a situation in no way requires us to abandon the goal of generalizing about social life, their arguments are not really ly persuasive (1971:132, 134).

Cybernetic theorists are tantalized by the thought that, somehow, cybernetic systems analysis could provide a way of reconciling under the single rubric of "process" subjective and objective social data.

32. For instance, in Kuhn's massive treatise (1974), there is no reference either to philosophers of language and/or hermeneutics (such as Ludwig Wittgenstein, John Austin, or Paul Ricoeur) or to sociologists of language, ethnomethodologists, and others who have made important contributions to concepts like "communication" and "transaction."

33. MacKay (1969); Krippendorf (1969a, 1969b, 1969c, 1971); Lin (1973); and Cherry (1966). This last book is a gem that no sociologist interested in communication can afford to overlook.

34. This is true both of Deutsch and Buckley.

35. See Dreyfus (1972); Hatt (1968); Crosson and Sayre (1967); Sayre (1969); Weizenbaum (1976); and Buckley (1968).

36. Education itself is affected by these tendencies as well as by cultural pluralism. Thus, the fragmentation of education destroys the integrating effects of superordinate visions of wholeness. This leaves the symbolic field clear for integral visions derived from the one agreed upon legitimating symbol—scientific technology itself. This is one of the points Ellul doubtless has in mind when he speaks of the rise of technology to sacred status.

37. Smith says these planners were not parochial (1970). But his meaning seems to be that they were not blindly politically partisan. They were surely parochial in their technological thought as evidenced by Smith's comments dealing with air force leaders' attitudes toward academic political science, the irrationality of base planning, and the ignoring of postwar international relations in exclusive favor of the goal of air force autonomy (ibid.:110–114).

38. My comments are informed by Sanders (1973) and Dror (1974). See also Dickson (1971).

39. In the English-language literature, these efforts are generally associated with the works, among others, of Harold Lasswell, Bertrand de Jouvenal, Yehezkel Dror, Bertram Gross, and Amitai Etzioni.

40. The conceptual analysis of the cybernetic versus bureaucratic society distinctions by Schick (1970) is an exception that lights the way to the work that needs to be done on that level. Several categories of empirical research are ignored by policy scientists (as well as many systems theory writers). Among these are the political sociology of information and expertise and the ethomethodological, labeling, and sociolinguistic styles of research into the communication process. Also relevant but generally ignored in policy science programmatic conceptualizations is the research literature that goes under the name of functionalism (not the more general sociological version but the more specific technocratic tradition in political science associated with Mitrany and his disciples). This literature constitutes a line of inquiry into the possibilities for redesigning social organizations according to functional, problem-solving criteria of social specialization. It is nothing less than twentieth-century empirical Saint-Simonism that has, ironically, been ignored by this century's would-be Saint-Simonist theoreticians in their plans for the redesign of the world system. See Mitrany (1966); Haas (1964, 1970); Sewell (1966); and Brenner (1969). Finally, there is the small but important research and evaluation literature specifically directed at the claims of technicist policy analysis. See, for example, Hoos (1972) and Brewer (1973).

41. Dror (1971b) presents his views on the education of policy scientists.

42. Note, by way of example in the empirical literature of policy analysis, the contrasting discussions of the Skybolt missile controversy in Sanders (1973:285–292) and in Steinbruner (1974:210–214, 234–237). Sanders, echoing Richard Neustadt's conclusions, presents the Skybolt controversy as resulting from "lapses in communications" and "major communication gaps both between and within the British and American governments." Steinbruner, on the other hand, interprets the whole matter with the aid of the cybernetic paradigm. In this context, communication lapses do not "just happen." They are accounted for by cognitive and organizational control mechanisms. In this light such controversies (and their causes) seem more inevitable, less accounted for by the "irrationality" of actors than by the "rationality" of cognitive and cybernetic mechanisms.

43. This bias against life's ambiguities seems a common feature of cybernetic technological thinking, to judge from the rhetoric of its partisans. It is muted among some, stark among others. Examples of the latter can be found in Helvey (1971: 7, 15–17, 20–21). On these pages there is a sort of animus against human natural languages, subjective values, human memory, and nonuniform sentence structures.

44. For example: "These arguments have been concerned with the how of governing, not with the extent of intervention" (Beer, 1974:45–46). "Thus the measure of the importance of cybernetics is the backwardness of society's present outlook on control, not the imagined ability of the science to build a super-brain" (Beer, 1964:207). Next to this, George is refreshingly direct when he says simply, "you have to believe that a system, such as a business, or a government, is like an organism and is capable of being controlled by a 'brain'" (1971:151).

45. According to Steinbruner,

Similarly, cognitive theory emphasizes that there is no guarantee to be found in reliance on seasoned men of good judgment. Though it is comforting to assume that the analytically bewildering complexities of policy problems are ultimately resolved in about the best fashion feasible when the decision process finally comes down to the exercise of judgment by responsible men, there is much in the foregoing argument to indicate that that is a bad assumption. In effect, the cognitive paradigm penetrates that often unexamined category called "judgment" and the results constitute a warning (1974:340).

46. Mikulak (1966:139). For a more extensive examination of this debate see Bakker (n.d.). Also see Graham (1967). For a description of the present role of cybernetics in the Soviet concept of international development see Ford (1966:169–192).

47. For a review of this tradition and its contemporary status in political science see Pateman (1970).

48. Compare this with the remarkably similar formulation in Helvey (1971:7, 15–17, 20–21).

49. These summary comments do not do justice to the imaginative details of Desmonde's theological effort (1971).

50. For Dechert all this rests on the assumption that "the relevance of this conception to the traditional Aristotelian-Thomistic doctrine of hylemorphism is clear" (1965:12).

51. Thus, Dechert tries to demonstrate the conceptual equivalence between theological and cybernetic categories. For example, see his discussions of teleology; of Providence and stochastic processes; of the soul; and of the correspondence theory of truth (1965:19, 17, 28–29, 27–28, respectively).

52. In his use of the word "beginning" Richardson does not make clear whether he means beginning in the chronological sequence or in the foundational sense (1967).

53. According to Richardson, "An intellectus is rooted not in thought but in feeling; and feeling determines the kinds of things about which we want to know the truth, the kinds of things that strike us as genuine problems. Contrariwise, an intellectus eliminates certain kinds of interests altogether by making them appear meaningless." An intellectus, then, is that "complex, unsystematic, pluralistic character of that matrix of meaning which is the felt basis for our discourse, not only about what is true and good but even about what is real" (ibid.:6).

54. Consider the large number of thinkers who have identified the red thread that runs through Western civilization, culminating in modernism, as the development of the cult of the person. For a fascinating and well written walk through this whole story, cf. Paul Zweig, *The Heresy of Self-Love*, New York: Basic Books, 1968.

55. It should be noted that this essay (1975) is based on a book on systems philosphy that Lilienfeld, as of this date, is preparing. It is not clear, and hence not fair to assume, that the author in fact dismisses systems philosophy in as single-minded a fashion as these excerpts suggest. I am using these quotations not prematurely to criticize Lilienfeld but to demonstrate a too facile dismissal of the significance of the systems philosophy movement. For the various claims and intentions of the systems movement as a general philosophy see the volumes by Ervin Laszlo and others under the rubric of The International Library of Systems Theory and Philosophy published by Braziller under Laszlo's editorship. I have not made reference to these volumes in this chapter, preferring instead to select examples for analysis from a wider range of literature. To do justice to the Braziller series would require a chapter in itself. However, I have studied them carefully and conclude that all my critical observations are, with fairness, applicable to them.

56. Note in the following passage how the rhetoric of anxiety can induce a mode of thought that lumps all sorts of intentions and actions of human agents, together with all sorts of classes of physical events, into a common category of "disturbances" that "impact" upon the servomechanistic cybernetic system.

> Disturbances can be classified as social, economic, political or natural. As an example of a social disturbance one can cite the revolution of rising expectations that is affecting not only our cities but the entire world. Recessions, wars, inflation, and high interest rates, which have a cataclysmic effect on the economy of the city, are economic disturbances which are beyond the control of a mayor. Political disturbances affect the city when changes in administration at the state or national level have a profound impact on urban programs. . . . For examples of government-induced disturbances originating at other levels of government, one has only to consider the highway construction and mortgage policies of the federal government, which "developed" the countryside surrounding the cities and peopled it with the cities' middle class. Also, it is evident that the nation's welfare policies, particularly as implemented in certain states, have influenced the rate of migration from the southern rural shacks to northern urban slums (Savas, 1970:1070).

This example of cybernetic thinking seems harmless enough, even trivial, when applied to the life of a city mayor. And, on the surface, we seem invited only to think in terms of standard functional relationships between social events. But what is the point of the cybernetic imagery then? Why is it added on to the catalogue of events we already know about any American city? The answer is that it occurs as part of an argument designed to persuade us to adopt this imagery as that of a technology of social control. If we take this invitation literally, we find ourselves imagining a city mayor, with immense resources of power in his hands, who continues to look upon his city as a would-be stable servomechnism and feels overwhelmed by all those external "disturbances" of his system dream. There takes shape an image of a man fully capable in good conscience of using his power to control or suppress these "disturbances" whose *noncybernetic* significance he is no longer able to comprehend. By a simple step of analogical reasoning we arrive along this road at a vision of cybernetic imperialism in which the strongest nation-states, in a Hobbesian world, use their power to control realities internal and external to themselves according precisely to these motivations. "Feedback" becomes internal surveillance and external espionage. And, again to quote Savas, in the name of "feedforward control," which "involves planning to accommodate predictable, externally caused changes that would otherwise impact the system," governments are made to fall, the armies of counterinsurgency are on the march in alien countries, trade is manipulated, and whole societies are reduced to pieces in a vast international cybernetic domino game. Such can be the euphemisms of anxiety.

PART III

Toward the Liberation of Practical Reason: The Problem of Countertechnicist Social Practice

INTRODUCTION

In this last part of the book, the time has come to move from the theory of technicism to the question of whether there are social practices that, in the name of rational freedom, can help counteract technicist corruptions of language as a form of freedom. From a moral evolutionary view of dignity, the question is: how can human consciousness be prepared for optimally rational personal participation in the evolution of the human world? From a more contemporary sociological point of view the question becomes: in a liberal democracy, can there be any institution whose essential mission is the practice of protecting the public language from technicist corruptions?

If it be assumed that consciousness is heavily influenced by early life experiences, then insofar as any social institutions could be said to be directly involved with the evolutionary fate of consciousness, they are surely the family and education. In a democracy, direct policy intervention into family life is morally suspect. Whatever hopes are expressed in a democracy for influencing the evolution of consciousness through the family usually are based on alleged indirect effects of education upon family life.[1] Insofar as it is assumed possible to speak of major public policies relative to combating the technicization of language in a democratic culture, then education is usually at the center of such hopes. Accordingly, in this part of the book I shall be concerned primarily with education in a democratic technological civilization.

There are two important ways in which the question can be posed of whether democratic education can have as its essential mission the protection of public language from technicist corruptions. These two ways flow from different senses of the phrase "can there be any institution . . . ."

First, one may ask about the empirical state of education as a social institution. What are the sociological trends in how education is formally discussed? What are the dominant official paradigms of educational policy questions? What is the relationship of these paradigms to the larger concerns of this book regarding technicist tendencies in modern society? In Chapter 7 I examine this set of questions as regards the institutional and official understanding of education in the United States today. The upshot of this examination is a demonstration that the logic of official educational policy thinking is, in subtle and interesting ways, quite compatible with a particular sort of technicist drift. It is likely to remain so, I intend to argue, because of a legitimacy crisis that now besets education as a publicly sponsored social institution.

The second way of posing my main question (i.e., can education be an institution whose essential mission is protecting public language from technicist corruption?) is this: is there any ideal conception of education, recognizable as part of traditional Western usages of the term, that envisions education as an essentially countertechnicist activity? If so, is this conception, viewed as a mode of democratic moral practice, compatible with the liberal political philosophy of individual subjective freedom? That is, would the practice of education as a countertechnicist strategy be innocent of the charge that it constitutes intervention into the subjective right of each individual to speak of the world in his own way? My general answer to this second set of questions is a qualified yes, and the grounds for the argument form the basis of Chapter 8. The fundamental qualification is that we have little sociological and ideological reason to expect that state agencies will comprehend education and educational policy issues in this essentially nontechnicist manner. I will argue that such a conception of education, although not incompatible with the logic of liberal democratic culture, is probably incompatible with liberal democratic state policy. This implies the urgent need for a democratic polity to regain its voice as a public independent of the state, capable of speaking about public policy as something more than a synonym for official state policy.

This thought, which ends Chapter 8, points in directions beyond the chosen confines of this book. Chapter 9 is a brief epilogue in which a few such directions for reflection are touched upon. Systematic pursuit of them must, of course, await another day.

Note

1. That such hopes are not entirely misplaced is suggested by evidence such as is found in Bowen (1977).

7 Education

THE CRISIS OF A CONVENTIONAL WISDOM

One of the more lasting clichés of American culture is that of the redemptive benefits of education. At present, partly because of the abstract exorbitance of past enthusiasms, this faith in education has fallen upon cynical times. Yet the cliché reflects a genuine, if conventional, wisdom. In Western societies, public education is an instrument of the polity's intention to address the young about their traditions, to induct them into the structures of citizenship and of useful activities, and to transmit the competencies necessary to comprehend and execute these activities. Faith in education should reflect a sense of the preservation of public traditions and a confidence in the elders' capacity to transform the young into creative participants in their perpetuation. A serious loss of faith in education presumably indicates a crisis of such confidence. Either the traditions no longer inspire moral commitment or they seem irrelevant to an inscrutibly complex future, or the young cannot be addressed as potential participants because the status of personal "participation" itself is in a stage of confusion.

It is proper, then, to examine public education as a potential intervention strategy against the technicization of modern culture. In this chapter and the next one I examine the implications of all that has been said for a sociologically informed philosophy of education. In this chapter I trace the growing legitimacy crisis of public education in a liberal, pluralistic democracy. Various policy proposals of recent years are reviewed. The point is to show that none of these contemporary proposals addresses the possible uses of public education as a mediator of competence for personal (not just functional) efficacy in a complex technological society. On the contrary, these proposals are consistent with the logic of that form of pantechnicism that I described earlier as a libertarian technicist order. The chapter ends with an account of what the minimum legitimate functions of public education are likely to be in such a society. Chapter 8 I take up the question of what a nontechnicist philosophy of education would have to be about.

Social Consensus and Public Education

In the United States, the legitimate mission of public education has always been ambiguous. This ambiguity is due partly to the inherent difficulty of defining education (as against cognate concepts like socialization, instruction, training, and indoctrination). It is due also in part to the ideological barriers to value consensus posed by a liberal democratic, pluralistic market society. Such a society is not, in principle, an explicit moral community. Rather, it is an association of individuals and groups integrated by economic exchange in pursuit of privately defined want satisfactions. Much of education, however, has evolved into a public enterprise. What, then, legitimates public education in the United States insofar as it deviates (as it always has) from instruction in commonly desired utilitarian skills? Whence comes the value consensus that ratifies the moral component of education as a public social institution? To deal with this question adequately, I must first look at some problems of defining consensus.

Clear definitions in social science are often achieved by models that simplify the complexities of real life. One such model is the notion of a perfectly integrated society. This model reduces to analytic insignificance the structural contradictions and linguistic ambiguities that generate conflicts in real life. This facilitates certain kinds of theoretical analysis. (Nineteenth-century anthropological thinkers sometimes confused such models with real primitive societies.)

In such an abstract model, consensus presents no problem. Society is merely reproduced (or "internalized") within the personality. Society being theoretically coherent and consistent, so is the model of the personality. The individual person becomes simply a "function" of the society. In such a model there is little variation between "inner" and "outer" experiences in life. According to the dominant sociological perspective on man, the emergence of a reflexive self (an "inner" person who questions, doubts, and chooses) is not a phenomenon given by nature in any direct way. Such a self is not born; it evolves in response to contradictions, conflicts, and ambiguities already present in the social environment around the self. In complex real life there are many such cleavages. Consensus is not automatic (as it is in a model of a perfectly integrated society). What can we mean by the term?

There have been at least two major types of answers to this question, one predominantly social psychological, the other more sociological. The social psychological type of answer ultimately locates consensus within the personality by postulating a psychic disposition (conscious or unconscious) called "attitudes." Consensus is then defined as common or converging individual attitudes toward a given issue. It is often very difficult, however, empirically to document convergency of this kind stable enough to account for a long-term historical phenomenon like an institution, much less a whole social order.

Sociological conceptions of consensus do not necessarily require such a stable convergence of individual attitudes.[1] Some sociological views of consensus lay their primary stress on socially structured power. The assumption is that, contrary to the attitude consensus argument, the facts of social structure are dominant. Those who control the facts of social structure are in a position to select the symbolic categories that play a part in any serious cultural debate and to direct the important decision processes that determine the outcomes of these debates. This, speaking generally, is the Marxist orientation.

There is another sociological approach to consensus that—although it recognizes the importance of power—does not make it the sole determinant. It is possible to argue that *a variety of determinants may converge* for a time in such a manner as to bring about a mass attitudinal support of an institution sufficient to legitimate it in public opinion. This is not a positive consensus, if one means by that term a collective, explicit agreement on the specific mission of the institution or social practice. (In terms of such a strict definition, there are probably few historical examples of true positive consensus.) Rather, I speak here of what may be called a "functional equivalence" mode of consensus. This phrase reflects a procedure whereby the sociologist looks for empirically verifiable structural and cultural features of a given society that appear to operate as substitutes for (or "equivalents" of) agreement of attitude among persons. It is important to understand that to claim that some institution or social practice serves as the functional equivalent to some other practice or institution is justified by a combination of theoretical and empirical arguments in the mind of a sociological observer. The equivalence need not be within the conscious awareness of the members of the society in question.

Let us consider a set of circumstances that can be so regarded as having functioned in this indirect manner, in place of a more direct value consensus among the American public on the moral ends of education. Roughly since the second world war, these circumstances have altered sufficiently to give rise to renewed confusion about whether a legitimate moral mission exists for public education. These alterations will be noted as well. Subsequent to that, I shall briefly review three educational policies that have emerged as proposed solutions to the legitimacy crisis of American public education.

The history and sociology of American education suggest at least six circumstances that in their combined effects could be thought of as having permitted the legitimation of public educational systems[2] in the United States. These circumstances are sketched in the following pages.

COMMUNITY AND SOCIETY: THE GREAT TRANSFORMATION

One of the dominant sociological themes of American history is its gradual transformation from an aggregate of communities to a complex national society. The impacts of this transformation upon American society and culture have been documented in every discipline.[3] This transformation is one of the major indirect causes of public education. Its effects operated indirectly by gradually removing the objective bases for the educative role of the family. The transformation from community to society was basically one of technological rationalization, institutional differentiation, and new interlocking patterns of coordination. Rationalization meant larger organizations and new technologically oriented job specializations. Institutional differentiation meant that new specialized institutions emerged to deal with many processes and problems formerly handled by local and multipurpose institutions like the family, the church, and the parish. New patterns of coordination spelled the decline of local decisionmaking contexts in favor of large cities, corporation headquarter centers, and the national government.

This transfer deeply affected the family and its functions. New opportunities for social mobility made family centered community and work life seem an irrational impediment to wider horizons. New social complexities and new technical skills made the family as an educative center seem parochial and obsolete. In general, the great transformation can be said to have created a niche for new educative settings in which the deficiencies of the family could be rectified with the aid of newly professionalized experts.

THE CONFOUNDING OF SECTARIAN AND MORAL EDUCATION

Historians have shown that Americans have always debated the moral meanings of education. However, sometimes it is instructive to look at the influence upon the outcome of such debates of the conceptual grooves along which the debate has moved.

One of the major grooves of such debates in American culture has been the question of church-state relations. There has therefore been a tendency to think of moral upbringing as an activity largely within the domain of the churches, leaving the so-called secular schools free to identify themselves with utilitarian ends as well as (once) noncontroversial moral themes like "good citizenship" and the "Protestant work ethic." The constitutional provision for the separation of church and state could thus appear as a basic solution to the conceptual task of separating moral and secular education, leaving the public schools free, so to speak, to specialize in the latter. Naturally, this solution is an illusion. The very term "secular" is a moral, if not a religious, concept. My point, however, is that when certain moral themes are taken for granted, they appear as so self-evidently naturally valid as to seem part of the general secular culture. (The confusion engendered by the virtual collapse of all such taken for granted assumptions about what constitutes secular consensus is perhaps in no way better illustrated than by the Conlan amendment.)[4]

THE CONFOUNDING OF EDUCATION AND SCHOOLING

Except in times of controversy, people normally do not inform themselves in detail about the actual educational content of schooling activities like curriculum, textbook selection, and administrative decisions. In an institutionally specialized society, these are considered the technical tasks of professionals, and people tend to be concerned primarily with the outcomes of these activities.

We may assume that a kind of tautology contributed to the acceptance of public education. *Schools*, once they were institutionally defined as nonparochial and therefore secular, became accepted as *educationally* secular by dint of that legal definition. The tautology is: what is education? Education is what happens in schools; what is school? School is where you go to be educated. This sort of tautology is a cultural feature of functionally specialized and differentiated social structures. Such tautologies inadvertently serve as a legitimating device for specialized role activities by short-circuiting what would otherwise be a natural tendency in a highly impersonal and specialized environment: asking everyone just what it is they are doing, how, and why.

THE UTILITARIAN CRITERION OF EDUCATIONAL OUTCOMES

In an environment of social specialization, people feel free to restrict their attention to the outcomes of institutional activities rather than worry about processes. Processes are regarded as safely delegated to specialists. We must not assume, however, that this concern with outcomes necessarily occurs on a high level of sophistication. It is more accurate to say that people are not normally very clear about just what it is they expect from any institution. In the case of schools it appears that outcomes have been assessed largely in utilitarian terms. These include the production of literacy skills, habits appropriate to business oriented social intercourse, plus motives and information useful for vocational advancement. Other aspects of the moral content of education simply received less attention. Such selective assessment helped to keep in the wings issues that could, in principle, be very controversial if placed directly under the spotlight.

THE UNIVERSAL MIDDLE-CLASS ETHOS

It has been shown repeatedly by sociologists that the concept of the "middle class" plays a powerful ideological role in American political culture. It symbolizes a variety of disparate values and motives such as assimilation, nationalism, upward social mobility, and general solidarity. It also seems to reflect false consciousness about the socioeconomic facts of life in American society such as the social distribution of wealth, opportunity, and power. People identify themselves subjectively as members of the middle class in the context of conditions that in many other countries would lead to corporate and hostile class consciousness.

This ethos must have contributed to the legitimation of public education. Indeed, it may well have provided the core content of whatever direct popular consensus there might have been on the moral mission of public education. Schools did present themselves as arenas for socialization into bourgeois aspirations, skills, and life-styles. To the extent that the American populace shared a general, if abstract, sense of the desirability of being "middle class," they accepted the schools as one way of insuring that their children remained in or entered that social status.

THE FACTS OF POWER

The foregoing influences obviously did not affect equally all sectors of the population. But they did not have to. The variety of social and physical controls summed up in the word "power" neutralize in any society a wide variety of potential dissidents. Those dissidents who lack political power to make their views potent can be ignored; they are, so to speak, politically invisible.

Throughout this century, the "real world" outside the schools that dominated the hidden agendas of school-based moral education was an industrial world. Its hierarchy, the basis for class definitions and for dramas of personal ambition, is a hierarchy dominated by business, managerial, and technological elites. The socioeconomic needs of a growing capitalist order thus provided an important incentive (and direction) for public education.

If these influences converged into an indirect pattern of legitimation for public education, why is there now such a storm of controversy regarding this institution?

The Legitimacy Crisis of Public Education

Many events, both social and cultural, directly and indirectly paved the way for the present legitimacy crisis of public education. Among the social events were the Great Depression, with its traumatic impact upon popular confidence in the self-regulating market, and the continued nonassimilation of some ethnic groups into the economy.

Among the cultural events, two are especially important. One is the general decline of the progressive faith—the confidence in technical and bureaucratically organized solutions to social problems. The other is the growing critique of the so-called Protestant work ethic. These and other events intersected during the 1950s and 1960s, resulting in a vigorous literature of social criticism directed against the established American order. This social critique can be generalized into three claims relevant to American education:

1. The American social order maintains its stability at the expense of a wide range of interests to be found in the population. These interests are of an ethnic, economic, racial, and sexual character. Public education helps to socialize children into this unjust order.

2. The institutional agents of public education represent the dominant, not the deprived, portions of the population both in their origins and in their orientations toward education.

3. The main rationale for public education, the utilitarian functions of schooling, has proven chimerical. Research suggests that schooling cannot be and never has been as effective a social equalizer as assumed by the public. Inequality in the primary arena of socialization, the family setting, appears to be far more potent than those who place faith in schooling imagine.

This conclusion, which bears on the genesis of public success in school, threatens to demolish the notion that *educational* policies designed to contribute to equality of opportunity can be successfully confined to issues of public *schooling*. If the state wishes to use public educational policies as a weapon against social inequality, it must be prepared to act in more radical ways. *Either* the state must move decisively in the direction of a planned economy designed to equalize the socialization platforms from which children are launched into the public education system, *or* the state must increase its intervention into the socialization activities of the family, the most private domain of society, in order to prepare children to take better advantage of existing opportunities for success in the industrial economy. Simply relying on public education to right inequality, some critics say, has even proven oppressive in some ways. The link that has evolved between hiring procedures and credentials based on levels of schooling is irrational and contradicts some of the civil rights of persons.

The mounting pressure of such criticisms has led to a general breakdown of consensus about the legitimate mission of the public schools. Even the general distinction between secular and nonsecular educational activities is now a matter of serious dispute. *All* the symbols and accoutrements of school activities (textbooks, tests, curricula, rules of etiquette, and even the teaching of standard, "middle-class" English grammar) have become objects of ideological dispute. A great debate is under way regarding the most fundamental moral aims of public education. Is a new consensus, or its equivalent, likely that is capable of underwriting a resolution of this debate through public policy?

What Is to Be Done?

Three types of arguments figure in current debates about public education and its mission. One is the call for an *explicit* consensus on a substantive set of moral ends for public education. The second is complete reversion to the market principle of consumer sovereignty; the third is total delegitimation of public schooling.

EXPLICIT CONSENSUS

Some commentators see the present discord as an opportunity to urge the public toward formulating an explicit consensus on the meaning and ends of public education. This seems an unlikely outcome of the present turmoil.

First of all, American society is undergoing an intensification of explicit value pluralism and its social correlates, a tendency that is incompatible with explicit national consensus on almost anything. In the face of this intensified pluralism, the major legitimating theme of modern public education seems likely to continue to be utilitarian vocational goals. Yet if further investigation bears out the research that suggests public schooling's inability to achieve this goal, then pressures will increase to delegitimate the state's direct interest in school activities on the ground that public schools are proving "irrelevant" to practical goals. In that event, industrial corporations will probably take over many tasks of explicit vocational instruction.

Sectarian educational philosophies remain important in American culture but, being embodied in private parochial school systems, they are not a factor in public education. If government financial support becomes legal, sectarian education will gain a new lease on life and may become a more significant competitor to secular educational assumptions. But even in parochial schools, education will not necessarily be governed by a cohesive sectarian consensus on the mission of parochial education. The varied motives that parents have for sending their children to parochial schools (such as avoiding racial integration) may abort any consensus on sectarian values on the part of parochial school boards. Furthermore, the state would probably set limits on the sectarian content of parochial education in return for financial support in order to remain within the bounds of the American Constitution.

The liberal arts education tradition has been gravely weakened by charges of "irrelevance" and by successful assaults against required curricula in the schools. The liberal arts tradition could become the moral specialization of the growing adult education sector in American public educational efforts since such programs attract many individuals who have come to experience dissatisfaction in the market oriented world of work. But certain features of adult education in America are likely to keep this possibility at the level of rhetoric. Among these features are the unsystematized contexts of the adult education sector, the underdeveloped state of pedagogy in relation to liberal arts education aims, and the persisting pressures of vocationalist and remedial schooling aims.

There is a further reason for the unlikelihood of an explicit consensus on the ends of public education, a reason rooted in an important characteristic of the modern concept of cognition itself. That reason is the educationally important metaphysical distinction between "factual" and "normative" domains of discourse.

In modern secular cultures, there is a strong intellectual trend toward distinguishing between the concepts of "fact" and "value." In the United States this is one of the important ways in which the doctrine of church-state separation is popularly interpreted. Public schools are not supposed to "preach"; they are supposed to teach respect for "facts." This cognitive orientation is likely to continue despite counterpressure from some quarters. The image of science as morally neutral knowledge is made credible by the fact-value distinction. Yet there is also increasing recognition that some moral socialization functions are in fact performed by the public education system. These contradictory pressures point to an intensification rather than a resolution of ideological conflicts in the school system. The sophistication of such debates will depend upon how well it is recognized by the public that the conflicts stem from contradictions in our entire cultural order. But such recognition is unlikely on a large scale. Social perceptions and definitions of problems are constrained by the blinders of institutional and role specialization. The agonizing problems of mutual unintelligibility not only between the partisans of "fact" and of "value" but also between professors of the "arts" and of the "sciences" have been glossed over by the rhetoric of "liberal arts education." It is becoming evident, however, that the potpourri transmitted under this label no longer constitutes a common framework of discourse that all educated persons can use to address each other on the details of important public issues. This, of course, facilitates recourse to technicist mystifications in public communication on education (as in all other issues).

Finally, let us glance briefly at the utilitarian approach to the problem of explicit consensus that is being promulgated in some quarters. Here is one example of such an argument:

> The way out of this problem . . . is once again to take the schools out of the business of making attitudes. Have them attend to skills, especially, in the beginning, reading, and the question of whose values control the schools becomes largely irrelevant. For it is my premise that the desire that children become functionally literate and able to understand mathematics is nearly universal; it is as true of poor as of affluent parents. . . . My point is that educational theory should define strictly educational tasks and that schools should concentrate on those. Any such definition must include, at one end of the spectrum, fundamental skills; at the other, it must exclude the conscious attempt to formulate social attitudes (Katz, 1971:143).

The explicit setting of educational aims in terms of specifiable skills is a reasonable strategy with which to turn public attention to the search for whatever common goals may be discoverable. But Katz seems to forget that "functional literacy," connected as it is with a person's most intimate relations with language, is a profound moral issue. There are no amoral skills.[5] This is why I will subsequently argue that the cybernetic metaphor, if it becomes institutionalized, will control (and partially reorganize) meanings of "utility."

CONSUMER SOVEREIGNTY

The intensification of the consumer sovereignty principle in the domain of education is quite marked today. We see it in the voucher plans, in the notion of learner sovereignty embodied in the proposals of Ivan Illich, in the awarding of "experience credits" and the proliferation of independent study plans in academia, and in the decline of consensus among college faculties on what courses should be required of all students. Where no agreement is possible, let the learners decide for themselves what their education is about. The principle of (self-defined) "needs" of learners is offered as the legitimator of institutional activity in education.

But this does not constitute a view of education. Consumer sovereignty is an abdication from, not a resolution of, discourse about the ends of education. As a policy, consumer sovereignty addresses the means of distributing resources, not the purposes of education. It is not even clear who the "consumer" is in such proposals. If it is the "learner," then what role have parents in the education of their children? If it is the parents, then what is

the meaning of the "learner's needs"? Moreover, formal education has traditionally implied the granting of access to values more universal in their
moral standing than the parochial claims of particularistic social contexts
(family, village, class). The consumer sovereignty principle defines education as a product fashioned to the given "needs" of the learner as the sole
moral point of reference. The principle thus actually delegitimates any universal values outside the particularistic influences that give rise to the immediate demands of the learner-consumer.

DELEGITIMATION

Some critics, most notably Illich, have argued for the delegitimation of
the public school system as a whole (1971). This, too, is no solution since,
as is evident in Illich's resort to the notion of a "learning web" (ibid.:72–
104), de-schooling society is sociologically equivalent to establishing consumer sovereignty in education.

Furthermore, de-schooling would contradict too radically the major sociological feature of modern society: structural differentiation. Public education is likely to remain institutionally specialized and separated in some
form from the broader patterns of socialization in society for the same reasons that it emerged as a specialized institution in the first place. Sociological evidence shows that this fundamental historical process of social specialization is accompanied by unavoidable tension. This tension is manifested
in all institutions via social problems associated with role conflicts, ambiguities of organizational prerogatives, bureaucratic rigidification, conflicts
over resource priorities, and inefficient redundancies of practice. Education
as a specialized social institution has obviously not been exempted from all
this. Some of these social problems may be expected to continue as long as
the process of structural differentiation continues to be a general fact of
social life.

There are still more basic reasons why neither delegitimation nor consumer sovereignty are solutions to the problem of how to attain consensus
on the moral ends of public education. These reasons are rooted in the ongoing transition from communal to societal organization.

Contemporary technological societies simply cannot be effectively organized any longer in self-sufficient, independent decisionmaking communities. Institutions are too interdependent, social problems too wide in their
applications, economies too large-scale, technology too integral to allow for
self-sufficient community life of the sort associated with previous centuries.
On the other hand, it is also true that we are witnessing communal reactions
against the enlargement of social scale.

It is evident that many of the specific problems of public education are
not strictly educational. They are facets of these broad tensions between the

communal and societal dimensions of modern social organization. Almost any significant efforts to resolve them will produce strains on the present constitutional order of the United States. The Constitution was simply not designed for the problems of late industrial society.

The fate of public education is intertwined with these broader questions. Any new kind of consensus on some *minimum* functions of public education will emerge as *part of* the struggle to resolve the larger constitutional questions of American society. This new sort of consensus may well be a libertarian technicist one.[6] A technicist consensus should be regarded as a profoundly moral, not just an amoral utilitarian, phenomenon.

Technicism is a moral issue because any consensus on problems of such macrosocial and cultural magnitude (including a technicist consensus) would eventually entail not merely ad hoc utilitarian compromises but, rather, far-reaching commitments to new moral meanings of the polity and, therefore, of civic life itself. Such new meanings must include reconceptualizations of education. Even in a libertarian technicist model of society, it remains necessary to presuppose some shared standards of competence for public life. However varied the moral frames of reference in such a society, however multitudinous the communal enclaves, some personal competences to adapt to a complex cybernetic environment must be universally inculcated. The nature of these capacities would determine the minimal legitimate socialization tasks of a national public education system.

Such tasks cannot be fulfilled by solutions such as explicit consensus on abstract values, consumer sovereignty, or de-schooling. The relegitimation of public education will evolve as the larger relationship between industrial society and modern culture evolves. I believe that the most plausible surprise-free projection for industrial democracies is the libertarian technicist model of development.

I shall now try to demonstrate the usefulness of this hypothesis for interpreting the "deep structural" significance of some specific policy proposals currently finding favor in federal educational planning circles.

Toward a Relegitimation of Public Education

To judge by the sheer volume of printed material, there is a good deal of policy innovation now going on in federal and local educational bureaucracies.[7]

For many of the educational policy enthusiasms of recent years it is tempting to accept as final James Koerner's interpretation:

> An irreverent observer who also is informed by experience might suggest that the support of educators for all of these dimly perceived ideas may have less to do with commitment than currency. A hundred million dollars or a couple of

hundred million Federal dollars is a powerful proselytizer. It represents a pot of money big enough to produce instant converts to almost anything (1973).

But other questions and possibilities intrude. There are patterns in these policy proposals and their rhetoric that may be clues to deeper sociocultural tendencies, in however latent a form. Since my concern is to provide one such interpretation, the temptation to indulge in extensive polemic about the worthiness of specific policies will be resisted. Instead, I begin with a brief account of three recent policy proposals. Then some common patterns are noted and interpreted in terms of my central arguments. The chapter concludes with a systematic depiction of a legitimate mission for public education in a cyberneticized, libertarian technicist order.

Among the many educational policy proposals that have emerged in recent years, three in particular have been receiving governmental attention: competency based education, career education, and tuition voucher plans.[8]

COMPETENCY BASED EDUCATION

This is a nationwide movement among some educators, administrators and psychologists to specify learning objectives more precisely so that what is to be learned can be disaggregated into interdependent components (or modules) that can be used in diverse situations with predictable outcomes. Partisans vigorously attack the practice of gauging educational outcomes in terms of cumulative time-in-classroom unit credit hours. Their argument is that such criteria have nothing to do with content; that they ignore learning outside the classroom; and that they fallaciously assume that classroom exposure *is* learning. Instead, the proponents of the movement stress flexibility in learning procedures (i.e., are willing to mix lecture, reading, individual programmed instruction through the use of computers, etc.) in favor of more attention to specifying what constitutes "mastery learning." They claim that this movement will encourage a renewed focus upon the aims and content of education rather than the structures within which it takes place. Furthermore, it is claimed that only such an analytic procedure can help to stimulate new educational methods capable of reaching the children of socioeconomically and educationally disadvantaged strata, who are entering college in ever increasing numbers.

A long-term historical perspective on this movement reminds us that competency based education is the latest stage in the application to education of analytic epistemology in general and the curriculum-packaging concept dating from the time of Peter Ramus in particular. More specifically, the competency based learning movement derives from the operations research procedures developed in World War II *and* from recent criticisms of the overbureaucratization of education.[9] The proponents of the movement argue that compared with earlier, more specialized notions of education, it is truly concerned with more holistic perspectives. This orientation results in some amusing formulations of "precise" educational competency criteria (e.g., "comprehension of aesthetic dimensions of knowledge," "proficiency in self-knowledge," and "comprehension of the relationships of man to his physical and social culture").

When one examines how such criteria get broken down into still more operational subcompetencies, one predictably finds that the easily measurable aspects of knowledge tend to be stressed. But the idea behind it all is accountability; the delivery on promises. "Thus," in the words of one partisan, "both the small and the large educational goals are considered, from the individual course objectives to the objectives of the institution itself."

CAREER EDUCATION

This movement can be understood in terms of one assumption and one goal that between them seem to sum up vast numbers of documents, pamphlets, and speeches. The assumption is:[10]

Where complex instructions and sophisticated decisions mark the boundary between the realm of man and the role of the machine, there is no longer room for any dichotomy between intellectual competence and manipulative skills and, therefore, between academic and vocational education.

The goal is:

All students should make a tentative career choice by the end of kindergarten and should modify or reaffirm the choice periodically throughout the school years.

Behind this lies the notion of career development:

The term "career development" refers to the total constellation of events, circumstances and experiences of the individual as he makes decisions about himself as a prospective or actual member of the work force.

There is a good deal of emphasis in the literature on a number of basic themes: the dignity of all forms of useful work; the more subtle motivations needed for sustained membership and mobility in the employment market; the importance of early-life career education as the basis for the rationality needed to steer oneself into a satisfying career; the need for *all* courses of study to be analyzed in terms of their possible career implications; and the distinction between vocational education and career education.

This last theme, the distinction between vocational and career education, is based upon at least two alleged differences. One is that although vocational education prepared the student for a specific occupation, career education (mindful of shifting specialization patterns) prepares him for a range of career "clusters" (e.g., business, environment, health, transportation). Career education is thus presumably much more flexible than its predecessor. A second difference is that although vocational education was treated as a separate curriculum from the academic track, career education is supposed to be completely integrated with the general curriculum. It is expected that this will restore the dignity of vocation and undermine the snobbish rejection by academics of career preparation as something fit only for rejects of the academic program.[11]

Opponents of career education, when they do not simply dismiss it as vague faddishness, condemn career education as

> a blinkered description of the purpose of education, a definition that is so uncompromisingly economic, so unabashedly narrow in conception, so relentlessly tied to the Gross National Product, and so anti-intellectual. (Koerner, 1973:11).

They also point to the fact that, for all the talk of the dignity of work, there is virtually no attention paid to substantive criteria of dignity or indignity. Proponents like to insist upon the importance of studying worker alienation, but these suggestions seldom are on a sufficiently macrosociological level as to bring into question the legitimacy of the total organization of modern industrial societies.

TUITION VOUCHER PLANS

There is a variety of these plans, but their basic purpose is threefold: the alleviation of bureaucratic centralization with its anti-innovative rigidities; the reintroduction of significant choice into parental decisions about their children's education; and the raising of educational quality by introducing greater competition into public education.

Under these plans the government would allot people tuition vouchers usable in approved, privately managed schools. Disagreements between proponents of different kinds of plans revolve around the fear of effects that would negate the main goals of general public educational policy, even given legislated guidelines for approved schools. Some of these feared effects are lowering the quality of education by neutralizing central regulation; redistributing public resources away from the poor and toward the rich; aggravating patterns of segregation (racial, ethnic, religious); increasing the possibility of educational fraud; and overstressing the "business" aspects of schools by forcing them to compete in the open market.

These accounts must suffice. My concern is not with the detailed history, description, and implementation of policies but with certain contradictions they share that are important to my central line of argument.

CONTRADICTIONS AND INTERPRETATIONS

At least two important contradictions seem implicit in these policies. They require some interpretation.

First, there is a clearly holistic (societal) rhetoric that pervades these proposals, but the proposals themselves focus exclusively on public education with minimal detailed attention to the larger context of that institution. For example, we find in these proposals strenuous efforts to combine rational manpower policies, early childhood socialization, and democratic values. We also find an ideological strain toward systems analysis and an opposition to specialized bureaucratic control over educational processes and definitions. (Note the tension in competency based learning between the established schooling notions of education and the idea of education as a universal dynamic process. Note the opposition of career education to the earlier conception of vocational education as a specialized activity. And note the threat posed by voucher plans to the hegemony of the established central educational bureaucracy.)

Yet for all this rhetoric, there is little truly societal analysis. For example, career education literature speaks of preparation for work but pays little serious attention to the likely employment patterns associated with future developmental stages of high technological societies. The industrial status quo is treated as a given, despite the talk of flexible career "clusters." Thus, we find little reference to the contradictions in the industrial status quo (e.g., the effects of persistent price inflation upon most people, who, however hard they work, have relatively stable or fixed incomes; the lack of fit between social needs and profitable market production patterns; and the increasing pressure on governments to be employers of first rather than last resort). The logic of competency based education *should* inspire a revival of interest in the educative significance of civic participation in all its forms; a theme of political philosophy stretching from the Greek polis to modern workers' self-management experiments. Again, no such connections can be found in the policy literature.[12] Finally, although the voucher plan debate does feature some of these considerations, even here the finance related implications have received precedence over others. This contradiction between rhetoric and analysis makes for an inevitably conservative bias in these proposals. A conservative bias toward the status quo may thus be the element that reconciles this contradiction.

The second contradiction that marks these policies is that one version of technicist argumentation is used to attack another version. Competency based education and career education entail reducing complex educational questions to extreme operational form for the purpose of accountability and outcome measurement; whereas the voucher plan policy reduces education to a commodity. Yet, as we have seen, all three policies are justified in the name of correcting excessive bureaucratic rationalization.

What looks like a contradiction is resolved by my comments earlier in this book (and by the analysis of Allan Schick, 1970) regarding the shift desired by some people from one technicist stage of social development to another. That is, these policies are reconcilable with the logic of an evolution from a bureaucratic (parochial technicist) to a cybernetic (pantechnicist) society. These policies reflect, on the level of educational policy, a transition from a type of thinking appropriate to a fragmented society of specialized institutions to an incipient, cybernetic perspective. In this cybernetic perspective, the interests of specialized bureaucracies are less important than the dynamic structure of the whole societal system. Institutionally specialized norms become less important than the universal reduction of all norms to operational terms appropriate for systemwide feedback analysis.[13]

Thus, career education is a partial recognition that future workers will not be monofunctional appendages to specialized machines but rather multicompetent tools in their own right. Therefore, they must be capable of being plugged in and out of a dynamic, innovative production technology. Competency based learning is a partial recognition that schooling is becoming prohibitively expensive. If more efficient investment and allocation decisions are to be made, better cost-benefit procedures must be established. In effect, the whole society must be treated as a vast educational resource. *This can happen only if educational ends can be precisely defined so that the educative dimensions of all life experiences can be harnessed instead of redundantly reproduced inside school buildings.* In this light, voucher plans may be interpreted as a recourse to market (supply-demand) criteria for rationalizing (and restricting) delivery of school-based educational commodities to precisely those consumers who really want and can pay for them.[14]

The voucher plans have another significance as well, one more directly related to the libertarian side of the libertarian technicist model. We have seen that according to this model societas and communitas will part company to the most extreme degree consistent with the survival of complex society itself. The moral criterion of society will be aggregate survival and all other moral phenomenologies will be relegated to the domain of privatized communities. In a libertarian technicist order, the state will have divested itself of authority over many kinds of moral definitions. What this means for education is that the public schools will indeed have to get out of the business of "making attitudes" about a wide range of issues. They will have to restrict their socialization functions to those that are minimally necessary to membership in a cybernetic society. All other ends of socialization (e.g., identity, roundedness, self-actualization, aesthetics, exotic modes of consciousness) can gradually devolve unto private systems of education. Voucher plans are, in this perspective, one experimental step toward a border politics of negotiation between societas and communitas.

One other policy should be noted, not in itself an educational policy but nevertheless one of great significance for our discussion. This is the very widespread trend in the United States toward *legislating* accountability criteria in public education. This process has been intensively studied, and the awesome conceptual confusion of these efforts are amply documented (Webster, Clasby, and White, 1973). But such confusion should not blind us to the significance of the effort and its scope in the light of this book's thesis.

I am now ready to raise the final question of this chapter. What would a legitimate public educational mission likely be in a frankly libertarian technicist society?

Public Education in a Technicist Society

There are at least three minimal socialization tasks appropriate to public education in any society with the properties of a libertarian technicist system. The ends of these socialization tasks are competencies that are essential for anyone who wishes to live effectively in society as well as community. Therefore, it is appropriate to make them the responsibility of public education. These competencies are the ability to speak across boundaries, the ability to help interpret the public interest, and the ability to participate in the public order.

A LINGUA FRANCA

A diversified social system requires a lingua franca, a mode of communication that people can use to span the system across all parochial boundaries within it. In a technicist public education, this superordinate language will be the positivistic vocabulary of quantifiable operational indicators. In a pluralist society, it is difficult to maintain any consensus on the validity of meanings that cannot be translated into operational terms of common utilitarian significance. (How troublesome it already is to speak of public values in an industrial society apart from formulas like gross national product.) We may therefore expect that mathematics, statistics, and computer related symbol skills will spread as a nationwide focus of curriculum revision and elaboration.[15] Conceived as a lingua franca, these vocabularies will settle the present ideological conflicts between partisans of "ethnic" and "standard" grammars. A lingua franca, by definition, is a language reflecting the existence of a pluralist environment. Children in a technicist society will learn it much as medieval children learned Latin in order to move across parochial boundaries.

PUBLIC INTEREST PARADIGM

A diversified social system requires symbols for representing the whole social order that its directors can use to legitimize actions they wish the public to regard as authoritative.

In a technicist society we would see efforts to move from an individualistic orientation appropriate for the liberal paradigm to the elaborated "public interest" implications of the cybernetic paradigm. Systems analysis has already considerably affected curricula in the social sciences and public administration. It would filter down the school system until the cybernetic metaphor became the basis for a new symbolization of the polity on all levels of public education.

CIVIC LITERACY

A complex society whose members wish to maintain a sense of democratic procedure must cultivate standards of citizenship skills appropriate to the degree of democracy they seek to maintain. In a technicist America, reliance on the democratic heritage of the Constitution is likely to continue, but with increased attention to the institutional levers for reinterpreting and updating its symbolic content (e.g., the general welfare clause). Chief among these levers, of course, is the Supreme Court, whose importance as a political invention is justly celebrated as part of the American political genius.

Technicism is compatible with an increasingly legalistic interpretation of the American civic tradition. With growing awareness of novel conditions never encountered by their ancestors, idealistic young Americans of the future technicist society may well cease to perceive the American myth in terms of a history of substantive shared ideals embodied in legendary great persons and events. Instead, they will see America as a system of rules reasonably malleable through legal adaptations to new conditions. This will inspire an educational aim of inducing a sort of mass legal virtuosity. "Popular participation" will come to be defined in practice not only as voting but also as active use of the skills necessary for harnessing abstract legal rules in the service of specific grievances or efforts toward reform. We already see this program being applied by partisans of the advocacy planning movement, Nader-style consumerism, the neighborhood legal centers movement, and the "welfare rights" movement (i.e., those who wish to use their legal welfare rights to overload the welfare system and thereby force basic reforms). It is a logical extension of such an orientation to argue that, these legal skills being the rightful heritage of every American and not just of an elite profession, efforts to transmit them should be centralized in the public education system.[16]

If such a program were fully successful, however, and unmodified by consensus on the appropriate bounds of legal coercion, the natural limits of what can be accomplished by means of legal techniques (as against other means of persuasion and other sorts of norms) would quickly become apparent. The entire society might begin to show the characteristics of a vast, static equilibrium comprised of veto groups at mutual legal standoff. At this point the ultimate crisis of the American civic consciousness would arise: the discovery that even the spirit of the Constitution has limits beyond whose boundaries lies the alien terrain of a more totalitarian technicism.

Such is the logic of contemporary American educational policy thinking. What would be required of education as an antitechnicist strategy? For this, we have to reflect upon education much more fundamentally, in keeping with an assumption shared by most great educational thinkers in the West: an adequate philosophy of education is ultimately rooted in a philosophy of experience.

Notes

1. Shils attempts to integrate the various theoretical orientations toward consensus under a common if somewhat eclectic purview (1968).

2. It should be noted that this legitimacy was never unqualified. The setting of public educational policy has never been an official federal function. Local prerogatives have been jealously guarded throughout American educational history. Hence, the plural term "systems" in the text is appropriate.

3. This is not a problem of recent origin. It is a major theme of American history. See, for example, Wiebe (1967). The tension between communitas and societas has also long been evident as a cultural theme in American literature (Marx, 1964). We also see historians of specific institutions relying on the theme of community-society transformation as in the case of Rothman's study of the history of the asylum (1971). For a general sociological treatment of the topic see Warren (1963).

4. The Conlan amendment to the National Defense Education Act, proposed on May 12, 1976, reads:

> No grant, contract, or support is authorized under this Act for any educational program, curriculum research and development, administrator-teacher orientation, or any project involving one or more students or teacher-administrator involving any aspect of the religion of secular humanism.

See the essays by Hook, Adams, Pfeffer, and others on some implications of this amendment in the *Humanist* (1976).

5. This issue is taken up further in the following chapter. Meanwhile we may note that research shows that many skills, including mathematical ones, imply whole moral settings within which they make sense. Some examples of such research are reviewed in Scribner and Cole (1973).

6. The reader is advised to reread my description of the properties of a libertarian technicist society on pages 39–42 of this book.

7. In 1968, two educational policy research centers were funded by the United States Office of Education, one at Stanford Research Institute, the other at Syracuse University Research Corporation (now Syracuse Research Corporation). I was for a number of years an associated research fellow of the EPRC/SURC. It is difficult to keep one's mental balance once one has plunged into the world of the usual bureaucratic educational policy documents or those emanating from the academic regions of the professional educational establishment. The features of such writings include momentary enthusiasms; a lack of historical sense (including any ironic appreciation for the fate of previous policy enthusiasms); hairsplitting over concepts of limited intellectual interest; almost anti-intellectual vagueness about really serious cultural, social, and historical traditions of analysis; and a technicist preoccupation with pseudoprecise "data" at the expense of attention to significant ideas about what, if anything, these data might signify.

8. The files of EPRC/SRC alone contain vast quantities of literature on these ideas, which yet form only a fraction of what exists. Just to give one example: for career education alone, one could obtain in November 1972 from EPRC Document Reproduction Service a seventy-four-page *Bibliography on Career Education* citing conceptual papers, journal articles, teaching guides, sample institutional materials, education reports, and research studies. As the basis for my comments I have consulted with some care the following documents on these three policy proposals: All these documents are available from EPRC/SRC.

COMPETENCY BASED LEARNING

Benjamin Rosner, *The Power of Competency-Based Teacher Education: A Report* (Boston: Allyn & Bacon, 1972); David McClelland, "Testing for Competence Rather Than for 'Intelligence,' " *American Psychologist* 28, No. 1 (January 1973):1–14; Harold Hodgkinson, "Issues in Competency-Based Learning" (keynote address to the National Conference on Competency-Based Learning, Cincinnati, February 15, 1974); Frederick J. McDonald, "The Rationale for Competency-Based Programs" (unpublished document by the chairman of the National Commission on Performance-Based Education); a series of five reports of preliminary case studies conducted by Thomas Corcoran for EPRC of Syracuse on competency-based experimental programs in selected academic institutions throughout the country; Harry S. Broudy, "A Critique of Performance-Based Teacher Education" (Washington, D.C.: American Association of Colleges for Teacher Education, 1972); plus other assorted unpublished memoranda and speeches. A very significant, nationwide study of competency based education reforms in American higher education is being prepared under the direction of Gerald Grant of Syracuse University, under the sponsorship of the Fund for Post-Secondary Education (forthcoming).

CAREER EDUCATION

Kenneth B. Hoyt et al., *Career Education* (Salt Lake City: Olympus, 1972); *Forward Plan for Career Education Research and Development* (Washington, D.C.:

National Institute of Education, Department of Health, Education, and Welfare, April 1973); *Abstracts of Exemplary Projects in Vocational Education* (Washington, D.C.: Division of Vocational and Technical Education, Department of Health, Education, and Welfare, June 1973); *Career Education: A Model for Implementation* (Washington, D.C.: Division of Vocational and Technical Education, United States Office of Education, Department of Health, Education, and Welfare, May 1971); *Report of AVA Task Force on Career Education* (Washington, D.C.: American Vocational Association, (1972); "Marland on Career Education, " *American Education*, November 1971 (Department of Health, Education, and Welfare reprint, publication number OE 72-52); "What Is Career Education? A Conversation with Sidney P. Marland, Jr., and James D. Koerner," Council for Basic Education, Occasional Papers Number Twenty (February 1973); Ross Burke et al., "A Preliminary Policy Analysis of Career Education" (Syracuse: EPRC Working Draft 7214, December 1972); "Comprehensive Career Education Models: Problems and Prospects" (Syracuse: EPRC Policy Memorandum 71-3, June 15, 1971); and Christopher Lucas, *Challenge and Choice in Contemporary Education* (New York: Macmillan, 1976), pp. 245–316; and Paul A. Olson, *The Liberal Arts and Career Education* (Washington, D.C.: Department of Health, Education and Welfare Monograph on Career Education, n.d.).

TUITION VOUCHER PLANS

Education Vouchers: A Preliminary Report on Financing Education by Payment to Parents (Cambridge: Center for the Study of Public Policy, March 1970); Irene Solet, "Education Vouchers: An Inquiry and Analysis," *Journal of Law and Education* 1, no. 2 (April 1972):303–321; and "A Proposed Experiment in Education Vouchers" (Washington, D.C.: Office of Economic Opportunity Pamphlet 3400-1, January 1971),

It is a notable feature of the arguments one finds in these documents that almost all partisans agree that no uniform, official definitions of these policies exist. However, I consider my description a fair and reasonable account of what these policies are about in the minds of their partisans. It should be noted that my interpretative focus is restricted to the macrosociology of culture. I am not concerned, for now, with the political determinants of these issues. Although duly acknowledging their importance, I think that preoccupation with the ever shifting shrubbery of politics can sometimes obscure the sociocultural forest.

9. I owe to Thomas Corcoran this insight that two separate trends, one technological and one antitechnological, have been captured, so to speak, within this single rhetoric.

10. The first passage is from "Vocational Education: The Bridge between Man and His Work, Publication 1, Highlights and Recommendations from the General Report of the Advisory Commission, Department of Health, Education, and Welfare/Office of Education (December 1967). It is quoted in Hoyt et al., *Career Education*, p. 60. This remarkable quotation comes as a startlingly precise affirmation, in a policy document, of cultural theories about technicism. The two subsequent quotations are from Hoyt, *Career Education*, p. 74 and p. 100.

11. This argument is made especially strongly in Olson, *Liberal Arts and Career Education*. The author is a professor of English who appears to believe in the possibilities of career education and places it in the context of a historical view of liberal arts education. This is no place to take specific issue with Olson's views except to say that they seem to me, although well intentioned, inadequate and naive when applied to state educational policy.

12. For another complaint along the same lines applied to contemporary political science theories of democracy see Pateman (1970).

13. The authors of these policies are usually not social theorists. But they are perhaps intuitively driven toward more systemic sociological insights by the changing structural circumstances around them. The conflict between their specialized bureaucratic loyalties and their intuition can appear as contradictions in their policy formulations. This is why simple debunking of these formulations is beside the point and uninteresting.

14. When schooling is legitimated according to the labor requirements of society, a distinction emerges between values directly relevant to this goal and other, "secondary" values that either may not be production relevant or may contradict industrial values. For an important discussion of implications of this distinction for Western discourse (both conservative and radical) on educational "equality" see Feinberg (1975).

15. Valuable comments on the moral significance of statistics and of operationist language occur in Sjoberg (1959) and in Arendt (1958:38 ff).

16. In this connection it is interesting to note in an article entitled "Study of Law No Longer Confined to Law Schools" that, as of the beginning of 1976, "between 400 and 500 elementary and secondary school systems have incorporated law courses into the regular curriculum, either as separate courses or as part of social studies programs" (*New York Times*, February 29, 1976, p. 9 of the News of the Week in Review).

Education

ON ENDS AND MEANS

We now face the task of exploring a philosophy of education inspired by a view of education as a strategy against the technicization of culture. My discussion is divided into three parts.

The first part has to do with the general question: what are the ends of education? In order to determine some answers, it is desirable first to distinguish education from other activities with which it is often confused. Here I seek to persuade the reader to a definition of education as an essentially antitechnicist activity, something not necessarily the case with the other activities I shall discuss. Then I go on to distinguish among the concepts of ends, functions, and utilities (often confused in policy discussions). After these preliminaries, I explore various ways of thinking about the ends of education.

The second part of our discussion of education emerges from these general considerations to focus upon the most problematic concept of education from the antitechnicist standpoint of this book. That concept is literacy. How a society defines literacy is no different, I argue, from how it defines the proper minimal relationship between every person and the sole universal instrument of world authorship: symbolization itself. I next compare technicist and nontechnicist conceptions of literacy. There follows a detailed, appreciative critique of the thought of Paulo Freire, the internationally known literacy specialist. I have chosen to examine Freire's thought closely—rather than John Dewey's, as might have been expected—for three reasons. First, Dewey's thought on education is more complex than Freire's and deserves a separate volume devoted to it from the standpoint of this book's concerns. Second, Freire has addressed the problem of literacy per se, in ways congenial to my own preoccupations, more directly than Dewey did. Third, Freire has become something of a cult hero among some human-

211

ist intelligentsias. This adulation is often rather uncritical, thereby obscuring some contradictions in Freire's thought that emerge when it is understood too simply as an ideology of revolution against cultural imperialism and social oppression. Noting both the relevance and the limits of Freire's achievements relative to the concerns of this book will help reveal some of the questions that a sociologically informed antitechnicist philosophy of education must consider in the years to come.

The third part of the chapter returns to the practical concerns of Chapter 7 on a different level. Here I explore the meanings of educational policy questions that flow from the philosophy of education set forth in the present chapter. These are policy issues as debated by educators conscious of themselves as educators not just as functionaries of bureaucratized and funded school systems. After some initial distinctions are drawn between different forms of the concept of policy in general, the discussion proceeds to educational policy issues appropriate for public consideration in a democracy. These are issues for public consciousness and debate rather than matters for technocratic state adjudication. Grounds for such a distinction are also considered in section three of this chapter.

Definitions

EDUCATION AND SOME RELATED ACTIVITIES

There are four concepts (socialization, training, indoctrination, and schooling) that are often used synonymously with the notion of education. It behooves me to distinguish them here in ways congenial to the intentions of my project. Since in real life these activities shade into each other, and since each of these terms is open to other than our definitions, the purpose of distinguishing them analytically in the manner I shall be doing should be stated at the outset lest the reader be tempted into suspicions of pedantry. I intend to argue in favor of regarding four out of five of these concepts as reconcilable with technicism, whereas education is an intrinsically antitechnicist activity in my view.

Why adopt such a controversial strategy? In order to refine and illuminate what education could mean if its essential end were regarded to be the development of persons immune to technicist mystifications. But why are the concepts of socialization, indoctrination, training, and schooling reconcilable with technicism whereas education is not? To answer this, I must venture some brief definitions.

Socialization is the sociologist's term for all those practices, both intentional and not, whereby a child is conditioned, taught, and otherwise induced to become a member of society, however membership is understood in any given society. Socialization is, by sociological definition, a universal phenomenon because it is necessary to the very concept of society itself as a historical (i.e., transgenerational) phenomenon.

I shall define *training*, in reasonable congruence with common sense, as a specialized activity whose end is the transmission of specifiable skills to a learner who has placed himself (or been placed) in the hands of the instructor for the explicit purpose of acquiring these skills.

Indoctrination, a controversial and complex concept,[1] I shall assume here refers to an activity whose primary end is imbuing individuals with an existing set of beliefs in the interest of perpetuating the authority of these beliefs.

Finally, let us regard *schooling* as an activity whose end is the transmission of knowledge in the form of thought modules called curricula.[2] Progress in schooling is measured by tests and cumulative certificates of grade completion.

Distinguished in this way from related concepts, education becomes a more subtle concept than would appear evident from its everyday use. Before venturing into this problem, let us say a word about what characterizes all the activities defined thus far in their common distinction from education. Socialization, indoctrination, training, and schooling all have as their ends the reproduction, within the individual subjected to these activities, of something. Thus, the end of socialization is the reproduction within the neophyte of dominant sociocultural norms; the end of indoctrination is the reproduction of some doctrine or belief; the end of training is the reproduction of some specialized portion of the social division of labor; and the end of schooling is the reproduction of ideas in the commodity form of packageable curriculum modules. Contrasted with these activities, education is something whose end transcends all forms of reproduction. I prefer to regard education as a process directed toward the cultivation of the person as a critical, reflective agent, aware of and capable of exercising intelligent freedom of choice over the reproductive interests of all systems of thought and practice. Why is this conception of education, not a new one to be sure, so persistently beset with complexity and frustration when it enters the arena of public policy discussion? As I have suggested, people often speak of education when they may mean activities that should rather be called socialization, training, indoctrination, or schooling. There is another reason, to which I shall now turn. People often speak of educational ends when they really mean what should be called educational functions and educational utilities.

ENDS VERSUS FUNCTIONS AND UTILITIES OF EDUCATION

To inquire into the utilities of education and its cognates is to be interested in technological means relative to clearly specifiable goals. It is a reasonable question to ask whether, when people speak of the utilities of education, they should be considered rather as thinking about the utilities of training. Most people probably mean little more by the demand that schools "educate" their children to read and write than what is implicit in the notion of training them to manipulate written symbols for the ordinary, pragmatic purposes of everyday life in a commercial society. People who regard the ends of education as I intend to discuss them should properly be very dubious about the uses of a phrase like "educational technology." Since this issue will reappear at the center of the second portion of this chapter, no more need be said here. It is relevant to add, however, that some people, when they use the term utilities, might usefully be interpreted rather as referring to the functions of education.

To inquire into the functions of education or its cognates is to be concerned with important observable by-products of educational and cognate activities. In such inquiry, one is interested in consequences as outcomes of structured activities quite apart from the intentions that give rise to these activities. Such consequences, or functions, derive both from the structure of the activities themselves and from the structure of the environment in which the activities take place. The functions of schooling in industrial societies, for instance, include keeping youngsters off the labor market and providing a backup socialization system for the family, a children's welfare system by way of school lunches and medical services, and "gatekeeper" social control through administration of public and private certification processes.[3] The disciplines most appropriate to the study of such functions are history and sociology.

The ends of education are neither utilities nor functions. What should concern those who want to speak of the ends of education, it seems to me, are the relationships between education and whatever one regards as worthy ends of personal existence in society and history. Such ends are not goals so precisely articulated that utilities can be technologically designed to fulfill them. Rather, such ends are better thought of as never fully specifiable criteria that one believes would fulfill one's deepest aspirations for personal and collective happiness.

To pursue this idea further, now that I have narrowed the scope of what "educational ends" will mean in this discussion, it is helpful to explore the general question of how best to think about educational ends. Not all approaches to this task are equally productive. The rest of this chapter reflects my prescriptive sense of how educational ends should be conceived, given the moral concerns underlying this book.

THE ENDS AND CONTEXTS OF EDUCATION

There are two popular but unproductive ways in which educational ends are thought about these days that are worth eliminating from our discussion at the outset. One is the tendency to promulgate an abstract rhetorical list of worthy "humanistic" ends, signed by a galaxy of supposed exemplars of these virtues. For reasons perhaps sufficiently self-evident for the moment, let us call this the "preachers fallacy."[4] The second unproductive approach is to accept as educational ends the goals implied by momentary popular enthusiasms, protests, bureaucratically funded rhetoric, or votes on education relevant referenda. We may call this the "plebiscitary fallacy."

Why are these orientations unproductive? Basically because they are irrational. They confuse ends, functions, and utilities, as well as education and its cognate activities. Proposals arising from the preaching and plebiscitary approaches tend to be abstract in rhetoric and insufficiently informed by theoretical analysis of the societal contexts to which they are intended to refer. For these reasons, they cannot set a rational standard for addressing the greatest single contemporary challenge to consensus on educational ends. This challenge is the cultural, social, and moral relativism arising from the complexities of modern society and from the collapse of general faith in a knowable, stable reality. The situation is well summarized by P. H. Hirst.

> If knowledge is no longer seen as the understanding of reality but merely as the understanding of experience, what is to replace the harmonious, hierarchical scheme of knowledge that gave pattern and order to the education? Secondly, there are equally serious problems of justification. For if knowledge is no longer thought to be rooted in some reality, or if its significance for the mind and the good life is questioned, what can be the justification for an education defined in terms of knowledge alone (1972:394)?

The situation in industrial democracies is now characterized by a pluralism of ends. These ends are arranged both "horizontally" (reflecting the interests of diverse institutions) and "vertically" in a stratigraphy of ideals that are the residue of past aspirations. In addition to clarity about what we want to mean by educational ends per se, there is need for some rational way to classify possible priorities in a relativistic environment. One way of doing this is to address the question of educational ends in terms of the various contexts to which they may have reference.

Obviously there are many approaches that can be taken to the definition of contexts. Let us examine four contexts that appear deeply relevant to educational ends: community, society, civilization, and crisis. These contexts are exclusive states, of course, only analytically. In reality they coexist, though their influence over the lives of populations are disproportionate at any given historical moment. I shall discuss each context as though it were a separate world posing its own unique challenge to education.[5]

Community. This is the world for which we educate when we are concerned with the cultural authority of the polity in its deepest sense. Community is the symbolic world of political tradition, myth, and story.

Education takes the form of preparing persons to assume their places as participating interpreters and authors of this symbolic world.[6]

Society. This is the world for which we educate when we are concerned with the domain of social utility. Society is the network of associations integrated by utilitarian rationality and the moral philosophy of contract: the world of technology, labor, and production. Education here takes the form of preparing persons to occupy their places as participating moral agents in a productive, socially organic division of labor.[7]

Civilization. This is the world for which we educate when we are concerned with the progress of universal reason. Civilization is the world of cosmopolitan norms of discourse, wisdom, adjudication, aesthetics, and standards of personal cultivation. For the great Western philosophers of education, education for civilization has implied preparation not just for appreciation of such norms but also for reconciliation of personal subjective freedom with civilization's objective development.[8]

Crisis. This is the world for which we educate when we believe anomic freedom is the dominant cultural fact of the times. This world is bereft of the authority of live public traditions. It is often characterized by terms like nihilistic, decadent, radically relativistic, or egoistic. It is a world racked with apocalyptic visions of its own end. Education here may take on a diagnostic and existential emphasis, designed to help create integral persons capable of sustaining moral projects in a nihilistic environment open to radical subjective freedom.[9]

Can we find implicit in these remarks a superordinate end of education as such? Education is, to be sure, concerned with the reproduction of culture, society, and civilization. But it is its essence that has been sought here by separating education from its cognate concepts, all of which concern reproduction. Education appears, in our reflections, as society's primordial obeisance to dignity. It is a concern for the cultivation of intelligent, responsible agency in a concrete world. This world may reflect, in differing proportions, the values of community, or of society, or of civilization, or even of the anomic freedom that is the natural soil of mythic creativity and renewal. What I mean is perhaps best communicated by two quotations.

> What concerns us all and cannot therefore be turned over to the special science of pedagogy is the relation between grown-ups and children in general, or putting it in even more general and exact terms, our attitude toward the fact of natality: the fact that we have all come into the world by being born and that this world is constantly renewed through birth. Education is the point at which we decide whether we love the world enough to assume responsibility for it and by the same token save it from that ruin which, except for renewal, except for the coming of the new and young, would be inevitable. And education, too, is where we decide whether we love our children enough not to expel them from our world and leave them to their own devices, nor to strike from their hands their chance of undertaking something new, something unforeseen by us, but to

prepare them in advance for the task of renewing a common world (Arendt, 1963a:196).

As civilized beings, we are the inheritors, neither of an inquiry about ourselves and the world, nor of an accumulating body of information, but of a conversation, begun in the primeval forests and extended and made more articulate in the course of centuries. It is a conversation which goes on both in public and within each of ourselves. Education, properly speaking, is an initiation into the skill and partnership of this conversation in which we learn to recognize the voices, to distinguish the proper occasions of utterance, and in which we acquire the intellectual and moral habits appropriate to conversation. And it is this conversation which, in the end, gives place and character to every human activity and utterance (Oakeshott, 1962:198–199).

Here we have it all: the nature of human worlds, the conditions of their survival, the continuity of generations. Education has to do with competent personal engagement with ancestors, contemporaries, and descendants in a conversation about traditions. Education is ultimately about time and immortality; about memory and testament; about all that makes of us more than creatures of biological destiny.

Literacy for a Culture of Conversation

Education as conversation: This thought sets the stage for the central question underlying these reflections on education. For a philosophy of education, what are the implications of technicism as diagnosed in this book? The dominant emphasis of this diagnosis has been a view of technicism as the mystification of personal potency and responsibility in relation to language. It follows that one crucial task for an antitechnicist philosophy of education is to set a standard for a conception of literacy appropriate to a society of *persons.*

It is remarkable that despite all the conventional wisdoms that have come under fire in recent years the notion of literacy itself has remained relatively immune from attack. Whatever else people fight about, it has been almost universally accepted as a good thing for everyone to be taught to read and write. The efforts of missionary societies, the United Nations, and national aid programs in mass literacy training are one of the few uncontested bits of conventional wisdom regarding the meaning of human progress. However, this sanguine view of literacy is something of a historical aberration. Serious thinkers about the nature of language and communication have long assumed that the acquistion of language skills is an act of profound moral significance, and the whole process has historically been associated with troubled reflection and controversy. From the biblical awe toward the world of God to the Sapir-Whorf hypothesis about language as the medium of ontology, the linguistic transmission of meaning has been regarded as much more than simply a problem in technique.

There is evidence now that this technicist attitude toward literacy as mere training is breaking down in some quarters. It is not that literacy as a valued end is being questioned. Rather, two questions are being raised. One deals with the meaning of the term literacy. Is it just a matter of knowing how to read and write? Is it that plus something else? Or are reading and writing perhaps not even necessary, much less sufficient, aspects of literacy. The second question pertains to the methodology of literacy training. Are the means of becoming literate of such moral significance that they constitute, in part, the ends of literacy? For example, if literacy means knowing how to read and write, is adequate literacy training merely a technical problem in how to teach people word recognition and reproduction in the shortest possible time? Or must it be assumed that a methodology of literacy training necessarily contains implicit within it a moral philosophy of education and society? What, in other words, is the wider moral and existential significance of mastering the process of "naming" the world in the myriad contexts of interpersonal dialogues?

My examination of this problem, properly restricted by the concerns of this book, will proceed along the following steps. I begin with a critical discussion of what seems to be a clearly technicist approach to literacy and to education generally. I then explore the scholarly lineaments of a nontechnicist approach to literacy that has been coming to life in recent years throughout the Western world. The third stage of my discussion is an in-depth meditation on the most nontechnicist possible sense of literacy: the personal capacity responsibly to "name" the world as a step toward world creation or destruction. This conception of literacy informs the lifework of Paulo Freire, one of the most influential practitioners of literacy education in the world; a cultic figure to many, an opportunity for debunking to others. In this section I examine Freire's thought in order to illuminate the ratio of insight and ambiguity that exists in the rhetoric of a nontechnicist conception of literacy. Fourth and finally, the chapter ends with some notes on foundational questions of educational policy implied by my reflections.

TECHNICISM AND EDUCATION

In the United States and fairly uniformly throughout the Western world, the definition of literacy is technicist in orientation. The United States Bureau of the Census defines illiteracy as "the inability to read and write a simple message in English or any other language" (Harmon, 1970:227). Perhaps the most starkly technicist definition of literacy to be found is the United States Army's notion of "functional literacy" set forth in World War II. Illiterates were defined as "persons . . . incapable of understanding the kinds of written instructions that are needed for carrying out basic military functions or tasks" (ibid.). Few definitions of literacy are as obviously technicist as the United States Army's. Indeed, that is its virtue. Essentially what it says is, you are literate if you can obey our orders when they are written down. The quantitative criterion was fifth-grade equivalency.

These examples suggest a pure technicist definition of literacy as follows: literacy is a *technique* (1) of sign interpretation and reproduction that (2) is, in itself, morally neutral to the same extent a tool normally is; (3) capable of multiple applications; and (4) operationally defined according to the utilities of sign interpretation and reproduction promulgated by those who consider themselves officially authorized to define literacy.

Literacy, however, can never be completely understood as a simple technique. The pivotal problem may be seen clearly as lodged in one word found in the preceding definition: *interpretation*. It is impossible to do without this concept since, whatever else reading is, it is irreducibly an example of interpreting signs.

Consider two definitions of literacy that open up a Pandora's box of complexities. In 1962, the UNESCO International Committee of Experts on Literacy adopted the following definition:

A person is literate when he has acquired the *essential knowledge and skills* which enable him to engage in all those activites in which literacy is required for *effective functioning* in his group and community and whose attainments in reading, writing and arithmetic make it possible for him to continue to use these skills towards his own and the community's *development* and for *active participation* in the life of his country. In quantitative terms, the standard of attainment in functional literacy may be equated to the skills of reading, writing and arithmetic achieved after a set number of years of primary or secondary schooling (quoted in Curle, 1964:12; emphasis added).

In 1970, a definition of literacy, not expressed in terms of grade equivalency, was advanced at a United States planning conference on the "right to read" principle: "The challenge is to foster through every means the ability to read, write and compute with the *functional competence* needed for meeting the *requirements of adult living*" (Harmon, 1970:228; emphasis added). As long as the italicized phrases remain obscure (as to criteria of content, reference groups, moral philosophies, and relevance to patterns of stability and change), these literacy definitions invite cooptation by elites able to enforce their own criteria upon people.[10]

The spectrum of disagreements that can result when people are invited to face such questions directly is well illustrated in the *Harvard Educational Review* issue on "Illiteracy in America" (1970). One finds here arguments ranging from pure technicism to outright attack on the notion that literacy should necessarily have something to do with reading. Given the theoretically unsystematic, ahistorical, and often ideologically flamboyant nature of these discussions, it was easy to feel some sympathy for proponents of the solidly technicist attitude embodied in this frustrated comment on the right to read goal:

> It's already seven days into the decade and I'd like to ask you all what we have done tonight to take one step forward toward that goal, if anything, except talk about getting involved and a lot about the theory of thinking and all that kind of stuff (ibid.:272).

Well-intended rhetoric is a no-man's-land between technology and philosophy. Either one stays on the side of operationally defined efficiency standards of literacy or one makes one's way to a larger view of what literacy means. In that case one finds oneself well beyond the confines of technical definitions; one is in the domain of social and political philosophy. As someone trenchantly put it at the Harvard symposium, "Do you know anywhere in the world of a mass education system that is not geared to socialize its people to the status-quo?" Sooner or later one must opt either against a status quo or for it and its possibilities. To do neither is to default on one's freedom of decision by becoming a literacy technician at the service of the highest bidder.

Why is it a profoundly technicist act to accept, without critical reflection, a social status quo as the stable environment to which education must adapt? Such a strategy is technicist because, whatever rhetoric to the contrary, education then essentially collapses into socialization. The reader will recall that earlier I identified the goal of socialization as the reproduction of society within the personality, with minimum emphasis on the development of the person as an intervening critic of society's claims. How then can we recognize a nontechnicist conception of literacy? We recognize it when a literate person is not regarded as one simply trained to accept a world presented to him as disembodied signals, superordinate goals, and managerial socialization. When, in other words, literacy is not treated merely as a utilitarian technique but as a composite of skills oriented to the achievement of a certain quality of personal consciousness.[11] It is time to explore further what this means.

NONTECHNICIST SENSES OF LITERACY AND EDUCATION

One of the characteristics of what I have meant in this book by a technicist culture is the mass displacement of a population's freedom and responsibility for action from the personal level to the technical level as represented by a society's technicians. Let us assume, without elaboration, the psychological generalization that when people, for whatever reason, feel no participatory responsibility for how their society is managed, they lose interest in the details of how it operates in fact. As society becomes more complex, whatever motives originally existed for technicist abandonment of personal responsibilitiy are intensified by the enclaves of secrecy generated by social complexity itself. These enclaves of secrecy (or "silence" if one thinks of secrecy as the withholding of communication) are generated both directly and indirectly. Enclaves of secrecy are the direct consequence of the myriad "security classified" types of information concealed by public and private governments in the interest of their struggles for advantage. Indirectly, enclaves of secrecy are generated by socially constructed mazes (red tape), and by proliferation of functionally *irrational* complexities (excessive "technicalities").

As these processes of mystification proceed, the public responds with its own forms of silence. That is, the public's resignation to the dictates of expertise and their subsequent withdrawal into the private hedonism of consumer existence seems accompanied by proliferating forms of refusal to endow their society with moral stature. Let us review some of these forms of refusal. Apart from the often noted escapist abuse of alcohol and drugs, they can also include *stylized cynicism* (refusal of esteem to anything); *voluntary ignorance* (refusal of hope that to know anything is to be able to change anything); *sabotage* (the refusal of respect for public authority implicit in the widespread motive to "rip off the establishment"); *child neglect and abuse* (in many cases reflecting a refusal authoritatively to represent society and its values to children); and *rhetoric of cultural revolution* (refusal of legitimacy to the mythological foundations of one's civilization).

The general relevance to education of such modes of public resignation is perhaps too obvious to require comment. The spreading revolt against schools, teachers, curricula, and canons of knowledge is evidence of popular antagonism toward a society whose moral credibility (or even sanity) is questioned. Here is not the place to pursue this interesting problem of the reciprocal modes of "silence" between persons and societies. They are of obvious diagnostic significance for understanding the condition of a moral order. I have gone this far in order to show what I mean by saying that literacy has to do with the dialectics of communication and silence between persons and their world. A world "speaks" to persons in the semiologies of objects (architecture, physical arrangements, fashions, artifacts, art products, etc.). It also speaks in the more systematic modes of language (speech, texts, music, etc.). To be literate is to "attend" to the world around us; to "interpret" what we hear and see; and to "name" in our own voices the conclusions that we are prepared to let inform our conduct.

Parenthetically, it seems to me unwise to force the concept of literacy beyond this point. Literacy has to do with the skills and arts of "conversation" between persons and their worlds. As such, literacy is dominated more by critical comprehension as an end than by successful conduct. Naturally, the line may not be drawn too sharply. To "name" the world with any adequacy is itself a mode of action, as the vigorous efforts of history's dictators to impose silence through fear and propaganda bear witness. Nonetheless, the concept of literacy should not bear the burden of defining all the criteria of practical wisdom. Life itself teaches us, at great cost, to avoid expecting facile continuities between intellectual and practical virtues.[12] For literacy to remain secure as an end of culture, the virtues of "knowing" for its own sake must be valued, even if any immediate hope of its translation into more direct forms of action is minimal.

In a moment I shall explore more concretely some of these matters in looking at the thoughts and career of Paulo Freire. But first, let us note some welcome signs in recent years of moves away from a view of literacy as simple technique toward one more congenial to the nontechnicist, agent centered conception depicted here.

There are diverse lines of inquiry, whose partisans are not always cognizant of each other, of which all those interested in a nontechnicist conception of literacy should take account. One such is historical-sociological research into the origins of communicative technicism.[13] Another recent development is a comparative anthropology of education more informed than former versions of such inquiry by awareness of philosophical problems in epistemology and cognition.[14] Third, there is the development of sociolinguistics. This line of inquiry has as its object the linkages between linguistic theory and ethnographic sociology of speech communities. In the minds of some exponents, sociolinguistics has explicit moral as well as scientific significance. At minimum, sociolinguistics studies linguistic agency for the purpose of placing persons back at the creative center of language. For some, sociolinguistics holds out hope for developing operational definitions both of competent communication and of equality of access to the resources of communicative competence. In this light, sociolinguistics is an important scientific component of what Habermas wishes to regard as the next stage in the moral evolution of reason: the development of a communicative ethics.[15] A fourth development of significance for a nontechnicist understanding of literacy is what, in some quarters, is called the "new sociology of education." This line of research is mostly a sociology of hidden agendas in educational practices. The focus of its major investigators is the social control effects of school activities and the functions of these effects for the socialization requirements of the social status quo. The import of these inquiries for literacy is that they do not take for granted the abstract rhetoric of educational intentions. Instead, they study the contradictions between what is professed and the hidden messages of linguistic and other actions on the part of official agents of education.[16] Finally, it is worth taking note of those various scholarly activities loosely identifiable under the rubric of hermeneutics. Some of these activities bear great potential significance for a nontechnicist pedagogy of the language arts. This is because the activities that go by the term hermeneutics seem to converge in the systematic search for insight, disciplined by relevant scientific findings, into what it means to interpret *any* text or text analogue.[17]

These separate lines of inquiry (historical-sociology of literacy, socio-linguistics, the "new" sociology of education, and hermeneutics) are some examples of work done in the last two decades that bodes well for the recovery of a nontechnicist meaning of literacy. Needless to say, we are far from seeing these inquiries affecting the established routines of university schools of education. But the very existence of these researches makes it possible to mount a critique of these established activities and to state what sorts of directions practitioners of education could be taking if they truly wanted to be more than technicians.

Even if all this were to occur overnight, what social changes would we have the reasonable right to expect? Not as many as some gurus of educational liberation like to imagine. For one thing, there is the question of how one avoids the critique of technicist literacy practices from itself becoming just another ideological rhetoric masking the interests of some new elite exercising its own will to power. For another thing, there are vast differences, as I noted earlier, between a "literate consciousness" and collectively orchestrated actions reflecting efficacious practical wisdom. To explore some of these issues (which is to say, to explore the limits of what we can expect even from the best intended education alone), let us reflect a moment on the works of Paulo Freire.

PAULO FREIRE: LITERACY AS AUTHENTIC DIALOGUE

It should be said in the beginning that Paulo Freire does not himself relate his views on literacy to the problem of technicism. Freire's rhetorical enemy is political oppression. For this reason some have been led to see his work as applicable only to societies like those of Latin America in which the processes of democratic liberalization have been aborted. His rhetoric sounds strange to ears that are accustomed to the unorchestrated demands of a rich multiplicity of interest groups. However, to dismiss Freire's challenge to modern education on these grounds would be a mistake.

Freire's assertion that a properly literate person is a person immune to political oppression involves a notion of literacy so fundamental as to transcend the single illustrative case of political oppression. As he himself says, the role of education in man's "ontological vocation to be more fully human" is what is really at stake. As Freire also recognizes, any discussion of freedom on this level must also entail the generic problem of "false consciousness." The deepest form of false consciousness is not simply passivity in the face of oppressive political mystiques; it is unquestioned obedience to the mystifications of one's own tools generally, be they economic, political, or physical. In other words, what profit is there in gaining freedom from the direct fetters of political oppression (as in Latin America) only to lose it to the more impersonal demands of technical efficiency and instrumental reason (as may come to be the case in the late industrial democracies)? It is questions such as these that make Freire relevant to this critique of modern technicism.

To understand Freire's work, it is first of all necessary to see what he regards as "literacy" and "illiteracy." After clarifying this I shall briefly discuss how he translates his philosophy of literacy into techniques of literacy training.

For Freire, literacy is a quality of consciousness not simply the mastery of a morally neutral technique. Since it is meaningless to speak of consciousness itself as neutral (i.e., undirected, qualityless), it follows that no aspect of language, an important mediator of consciousness, can be regarded as lacking moral significance. Freire draws from this the inevitable conclusion: "Every educational practice implies a concept of man and the world" (1970b:205). To learn, as to teach, is an act. Freire's discussions of this point reflect awareness of the influence of instrumental reason upon contemporary conceptions of consciousness. He seems well aware of the subject-object problem as it pertains to education.

> This process of orientation in the world can be understood neither as a purely subjective event, nor as an objective or mechanistic one, but only as an event in which subjectivity and objectivity are united. Orientation in the world, so understood, places the question of the purposes of action at the level of critical perception of reality. . . . Men have the sense of "project" in contrast to the instinctive routines of animals (ibid.:206).

NAMING THE WORLD.   From this concern for the proper understanding of subjectivity and objectivity arises Freire's persistent attacks upon what he calls the "banking" conception of education: the notion of learning whereby the learner simply digests knowledge, being "passively open to the reception of deposits of reality from the world outside."

> Implicit in the banking concept is the assumption of a dichotomy between man and the world: man is merely *in* the world, not *with* the world or with others; man is spectator, not re-creator. In this view man is not a conscious being . . . he is rather the possessor of *a* consciousness. . . . This view makes no distinction between being accessible to consciousness and entering consciousness (1970a:62).

It follows that there is no such thing as a value-neutral literacy training text or method. All words either conceal or reveal something. Perhaps the essence of Freire's conception of literacy is found in the statement: "Teaching men to read and write is no inconsequential matter of memorizing an alienated word but a difficult apprenticeship in naming the world" (1970b:211).

This point, of course, raises the question of whether there are valid and invalid ways of naming the world. Freire obviously believes there is a valid way: "The role of the problem-posing educator is to create, together with the students, the conditions under which knowledge at the level of the *doxa* is superseded by true knowledge, at the level of the *logos*" (1970a:68). What does he mean? Freire seems to mean two things by *logos* in this context. One is a "demythologized" awareness, essentially a man's awareness of the true state of his position in the socioeconomic structure in which he is situated. The other meaning of *logos* to Freire is more philosophical in tone: a man must understand his "ontological vocation to be more fully human" (ibid.:61). This conception is never explicated clearly in Freire's writings, but it would seem to refer to man's exercise of his freedom to name the world. For Freire, the phrase "to name" has a creative and transformative (world constructive) connotation. All those who either knowingly or unknowingly stand in the way of the progress of humanization in these terms constitute, in Freire's vocabulary, "the oppressor."

THE ILLITERATE.   On the basis of what has been said of Freire's conception of literacy, we can readily understand his idea of who the illiterate is. First of all, illiteracy is "a typical manifestation of the 'culture of silence' directly related to underdeveloped structures" (1970b:209). The illiterate is normally viewed as on the fringe of society, as marginal to the socioeconomic and political order. Often he is interpreted as being there by choice—as resisting literacy or not having the will to learn.

> In accepting the illiterate as a person who exists on the fringe of society, we are led to envision him as a sort of "sick man," for whom literacy would be the "medicine" to cure him, enabling him to "return" to the "healthy" structure from which he has become separated. Educators would be benevolent counsellors, securing the outskirts of the city for the stubborn illiterates, runaways from the good life, to restore them to the forsaken bosom of happiness by giving the gift of the word (ibid.:211).

Freire rejects this concept of marginality out of hand. Rather than "being outside of," illiterates are "beings for another." That is, they are unconscious of the internalized oppressor within them. This internalized oppressor is a composite of the assumptions about the poor and the powerless held by those who rank high in the social structure and reap its benefits. In many cases, the powerful and the powerless share a mythologized sense of the social order as in the "nature of things"—a natural order of superiors and inferiors. Thus, the illiterate, among the most powerless and helpless of men, does not realize that "men's actions as such are transforming, creative, and re-creative." Therefore, literacy training will simply not take hold, will not engage men's capacities for action, unless such training relates "speaking the word to transforming reality" (ibid.:213).[18]

Such is the philosophic basis of Freire's approach to literacy. He has developed a series of technical principles for literacy training whose rationale is grounded in this perspective. The overall purpose of these techniques of "problem-posing education" is the achievement of "authentic dialogue" between educator and learner. By this term Freire means dialogue that enables the illiterate, in terms significant to him, to discover the oppressor within him, to cast off his "mythical" false consciousness, and to achieve his "onto-logical vocation" as a creative agent. This agenda presupposes stages of preparation.

GENERATIVE THEMES. The first stage is "thematic" investigation. By this term Freire means intelligently conceived and executed anthropological research on the objective situation and the symbolic world of the illiterate population being worked with.

> The task of the dialogical teacher in an interdisciplinary team working on the thematic universe revealed by their investigation is to "re-present" that universe to the people from whom he first received it—and "re-present" it not as a lecture but as a problem (1970a:101).

The translation into a problem that the illiterate can understand (including his own place in it) proceeds by way of collecting a number of "generative words." These are words that, having been derived from within the illiterate's own symbolic world, lead him to ask an expanding series of questions about his situation. Gradually, "generative themes" emerge as the illiterate becomes literate and able to put words together into more complex combinations of ideas. Together, always in dialogical manner, educators and learners move through the literacy materials (words, texts, pictures), gradually "de-coding" them. These literacy materials, or "codifications," are not "slogans; they are cognizable objects, challenges towards which the critical reflection of the decoders should be directed" (1970a:107).

Ideally, codifications are organized as "thematic fans"; that is, in mosaics of themes that suggest each other in multiple possible patterns. The learner is progressively inducted into awareness of the contradictions that constitute his societal environment, culminating in his comprehension of how he—a creative agent—was made a "being outside of" society.

Informing all technical considerations is Freire's underlying perspective on a free man's ontological vocation. Thus, he goes to considerable lengths trying to distinguish between truly revolutionary dedication to "the people" and authoritarian teaching; between propaganda and codification; between manipulation and education; and between true and false revolution. For Freire, any meaning of literacy short of liberation in these terms constitutes not social development but cultural invasion.[19]

A CRITICAL COMMENTARY ON FREIRE

Freire's attack upon the established notion of literacy poses a challenge difficult glibly to dismiss. To be told that one's ways of educating produce people who are stunted human beings, helpless before an inscrutable though manmade environment, strapped into unnatural strata of power and subordination, and mystified by institutionalized mythologies is to feel called upon for a reaction. And if one can succeed in demonstrating that the most powerless and marginal of persons (functionally illiterate peasants) can be inducted into a life of responsible agency, then one has cause to believe it can be achieved with anyone, despite the celebrated failures and costs of formal educational efforts. Freire's work is not only an indictment of established educational practices, it is also a vote of confidence for what can be called the humanistic possibilities of correct methods of educational intervention into even the most degraded circumstances of social oppression and false consciousness.

Given such stakes, it is obviously important to develop ways of assessing the arguments of radical educational thinkers such as Freire. If carefully articulated, such assessments can assist in the formation of a sociologically informed philosophy of education in societies that stress the dignity of the person as a responsible agent.

There seem to be two fundamental lines of criticism that need to be explored regarding Freire's educational thought. They can be introduced in the form of questions. First, does Freire assume too simplistically that some precisely specifiable alteration of a given social practice (like literacy training) will resolve the moral problems defined in his critical diagnosis of existing society? Let us call this the problem of utopianism. Second, is Freire's argument adequately elaborated in terms of its practical implications for the social order at large? Let us call this the problem of unelaborated social consequences.

THE PROBLEM OF UTOPIANISM. Utopianism is, in my judgment, a drawback in Freire's thought. It is evident in an uncritical tendency to regard his notion of literacy as the key to liberation and a life of rational action for all persons. This is to say that he does not apparently take much note of the complexities, much less the dark side, of the notion of liberation itself.

Let us enlarge on this point by way of three commentaries. I shall discuss the ambiguities of liberation defined as men's realization that they make their own nature and their world; the issue of latent elitism in Freire's thinking; and the problem of chiliastic impatience.

The Ambiguities of Liberation. There are at least four quite different ways of asserting that men determine their nature and their world. Each of these ways of thinking about liberation possesses moral implications and problems unique to itself.

First, one may say that men collectively invent their conduct and generate the coordinates of the human world (language, arts, science, artifacts) but are helpless to do so in a deliberate way according to plan. One contemporary exemplar of this view is Leslie White, (1949) who insists upon a discrete discipline of "culturology" largely because he regards culture as a transgenerational causal force superordinate to the intentions and desires of people in any given generation. For White, awareness of the causal role of culture in human existence may legitimately be regarded as an important source of insight into historical changes and hence as a resource for more intelligent adaptation. However, White scorns the view that men can control the evolution of their civilization in any way they please at a given moment (ibid.:330–359). In the context of such a view, revolutionary action for the sake of liberation makes little sense. An educated person is one who maintains optimal critical awareness of the relationships between cultural evolution and the conditions in which he finds himself.

A second way one may assert human determination of self and world is to say that men invent their world through trial and error. They do so by way of processes that—if people are left alone to pursue their interests rationally—are more benign in their results than could be said of any product of deliberative design. This is the central faith of Anglo-Saxon utilitarian liberalism. To those who share it, rational action is possible only in the context of a free market (consumer sovereignty in economic production and political choice). Inevitably one is then heir to the well-known tensions that exist between the values of liberty and equality, of which Alexis de Tocqueville was such an acute prophet.

A third interpretation, one of rather tragic grandeur, is the view that men create their world only in rare moments of true action under the impetus of charismatic inspiration. With the routinization of charisma that inevitably follows, the logic of the factual world, so to speak, takes on its own impersonal way once more. Such a view, exemplified best in modern times by Max Weber, lends itself to a notion of action as clearing the way whenever possible for fresh episodes of charismatic renewal. In the name of this abstract goal, one could advocate in practice anything from resistance against excessive bureaucratization to frankly romantic rejection of modern rationality, pending the advent of new Caesars or messiahs.[20]

Finally, it could be argued that men are destined one day to evolve into some sort of "true" freedom whereby they consciously and cooperatively appropriate the world and transform it into a structure designed for optimal personal fulfillment. As phrased, this is an expression of Marxist humanism that Freire's life in part reflects. But, as I noted earlier in this book, from Henri de Saint-Simon to modern policy science, a less romantic and more elitist technocratic version of this ideology also exists.

It thus appears that literacy, simply regarded as awareness that men can make their world, is an insufficiently explicated legitimation of a progress oriented literacy training.

Latent Elitism. Having reviewed the first problem pertaining to Freire's understanding of the relationships between literacy and action, I turn now to the second: the possibility of latent elitism.[21]

The possibility of latent elitism hovers behind any theory of false consciousness. He who believes himself to understand the dimensions of true consciousness, whatever they are, must necessarily—however temporarily—play the role of an elitist guiding the unenlightened to their proper destiny. Freire's views, as we have seen, place an extraordinary emphasis upon education as an instrument of liberation. If Freire were to carry the matter further and admit that social mobilization of large numbers of unenlightened people is also necessary for revolution, a sort of Leninist elitism would have to be the next step in his thinking. Under such conditions of mass mobilization, both church and secular history suggest that the saintly educators on whom Freire depends to keep his revolution honest would turn out to be in short supply.

More fundamentally, the risk of unadmitted elitism stalks anyone who insists on using the term "myth" to characterize perspectives with which he disagrees. No faith perspective on what it ultimately means to be fully human is any longer completely immune to the challenges of cultural relativism. In other words, there is no way in which intellectual criteria alone can be used to demonstrate conclusively the timeless validity of one set of moral axioms against all others. This understanding has led to a new respect for the role of myth in social life.

It is, of course, likely that Freire is aware of such problems and would apply the term myth only to self-evident justifications of social oppression. Freire's work and life as a whole show ample desire to walk the delicately visible line between paternalism and relativism. Nevertheless, it is also true that there has been a conspicuous failure on Freire's part to attempt a more universally relevant definition of "oppression" in his writings. As it stands, his approach to this concept is dangerously abstract. Somewhat special circumstances characterize social stratification in parts of the world such as Latin America, which Freire has in mind. Application of his simplistic notion of oppression in the developed Western world and some other parts of the Third World could have tragically irrational consequences. Extreme circumstances (e.g., ignorance, helplessness of peasants, internalization of ruling class contempt) that might justify an all-out revolutionary populism simply are not equally distributed throughout the world. Yet Freire writes as if they were. Furthermore, Freire's very definition of oppression as "any situation in which "A" objectively exploits "B" or hinders his pursuit of self-affirmation as a responsible person" (1970a:42) skirts dangerously close to the totalitarian doctrine of the "objective enemy," which Hannah Arendt so rightly condemned.

> The chief difference between the despotic and the totalitarian secret police lies in the difference between the "suspect" and the "objective enemy." The latter is defined by the policy of the government and not by his own desire to overthrow it. He is never an individual whose dangerous thoughts must be provoked or whose past justifies suspicion, but a "career of tendencies" like the carrier of a disease (1958a:423–424).

Some of Freire's terminology, then does lend itself to charges of elitism and even of revolutionary totalitarianism. My defense of his intentions should not obscure my sense of Freire's failure, as a *thinker*, to elucidate better the problem of how to distinguish elitism from redemptive dialogue.

I have now reviewed critically two facets of Freire's thought as regards what was called his incipient utopianism. One is his failure to appreciate the conceptual and moral ambiguities of the idea of liberation. The other facet is the potentiality for elitism in his rhetoric. Let us now turn to the third facet of his strain toward utopianism.

Chiliastic Impatience. To put it briefly, Freire seems to place most of his hopes for the future on educational enlightenment about the true nature of human agency. Nowhere does he seem to consider the possibility that a significant proportion of people might come to reject such radical freedom in favor of benign authoritarianism and aesthetically tinged mystification. Yet it would not be the first time people did this once they discovered that real personal autonomy meant hard work, unrationalized frustration, and "too many evenings." Radicals generally are not wont to consider with sympathy the arguments made by Dostoevski's Grand Inquisitor. Yet, as much evidence surely exists for this interpretation of human motives as for any other. From this perspective, intellectuals like Freire, who refuse to cease their quest for the Holy Grail of liberation, could be regarded as purveyors of a socially dysfunctional chiliasm that corrodes like acid the stability of achieved moral order.

It may seem almost vulgar to bring forward such considerations under the conditions of Latin American and many other societies. But Freire aspires to a universally relevant philosophy of literacy education. And, in the name of his vision of liberation, he tends to downgrade revolutions energized by lesser aspirations. Is it not necessary (even perhaps in very oppressive societies) at all costs to avoid dividing the population into airtight compartments of "oppressor" and "oppressed"? Such thinking induces expectations of some utopian Armaggedon that will purify the realm of all versions and vestiges of the ubiquitous "oppressor." In such an artificial polarization, most of us incapable of some masochistic renunciation of our traditional identities might well find ourselves in the "oppressor" class. This completes my first line of criticism: Freire's utopianism. Now let us turn to the second.

THE PROBLEM OF UNANTICIPATED SOCIAL CONSEQUENCES   Freire does not go far toward exploring the practical implications of his literacy program for the larger orders of society and culture. Without benefit of such sociological extrapolation of possible consequences, there is much risk in heeding the call to cultural revolution.

I will not spend much time with Freire on this topic because his attention is focused on the issues immediately surrounding adult literacy training. However, such refinement is necessary for Freire if his ambition is the development of a universally relevant view of literacy as a determinant of social change. Three very brief reminders of needed refinements must suffice here.

First, if carried to its logical conclusion, Freire's notion of literacy as a determinant of social change requires a virtual commitment to social development as persisting revolutionary destabilization. This is because not all sectors of the population can simultaneously be made literate in Freire's sense. People's capacities, experiences, interpretations of life, and ambitions differ too much for that. Furthermore, it is very likely that people's responses to such a form of "awakening" will vary. They will range from utilitarian egoism through communalist fervor to apocalyptic enthusiasm. In brief, Freire's sense of literacy will not mean the same thing to all people. Freire seems to assume that it will. He has nothing to say about how such profound energies, once liberated, are to be structurally channeled so as to avoid both the destructiveness of persisting destabilization and the recapitulation of past forms of oppression.

Second, I have already commented upon the dangers of applying a simplistic oppressor-oppressed dichotomy even to heavily caste stratified societies, much less to all societies. Yet this dichotomy plays an important role in Freire's literacy theory. One cannot help wondering whether Freire, from a social structural point of view, may not be inviting more opposition by way of some of his rhetoric than his ideas warrant. Finally, the notion of a fundamental "new man" literacy is itself not new. The task for a radical analysis of literacy is to evaluate the religious and secular history of similar efforts in the past. Such studies, combined with evaluations of contemporary efforts, can stimulate a much more refined analysis of the redemptive powers of literacy under varying conditions.[22]

For this to happen, however, more unashamed confrontation with the elitist implications of new man literacy doctrines is necessary. History does not leave us sanguine about the efficacy of Freire's call to "revolutionary trust" in the people. On the contrary, it rather seems to prepare us, in Carl Bereiter's words, for "something of a paradox: the most important educational goal of free men is one that men will not pursue of their own free will" (1971:47). One may, of course, respond by simply asserting a contrary faith, as Everett Reimer does:

> I believe that a proper organization of society could make available to everyone the circumstances that only scientists now enjoy and that under these conditions everyone would find the pursuit of reality natural and enjoyable (1971:89).

But it is surely not unreasonable to ask that one prepare to be wrong. In that case, one would have to choose between relativism and elitism as avenues of desired social change. Such choices need sociological elaboration of a sort that Freire nowhere provides. Not to provide it is to invite having one's methods reduced to simple techniques in the service of quite different intentions.

In this section, I have articulated views on the universal significance and existing limitations of Paulo Freire's thought on literacy. The efforts of a sociologist to contribute to the cumulative historical conversation on education would be insufficient without some attention to the enduring questions that must be addressed by anyone wishing to enlarge the meaning of literacy explored in this chapter.

Policy Issues for a Nontechnicist Educational Philosophy

In Chapter 7 I criticized current American educational policy discourse as having little to do with education. Now is a good moment to integrate these reflections by asking what, in the light of this present chapter, the questions of educational policy are really about. First, however, a little stage setting might be helpful regarding the term policy itself.

Most generally speaking, policy is organized conduct oriented to commonly valued ends. Policy is morally informed and intentional conduct, not simply technologically rational behavior relative to precisely specified goals. Policies, in short, are not merely techniques. There exists a paradox in contemporary American life with respect to public policy. On the one hand, there is everywhere talk of public policy. The several sciences are increasingly judged in terms of their policy relevance; government is being called upon daily to "solve" social problems; and policy study Institutes are proliferating in academia. Yet, never has there been such a massive public reaction against governmental intervention into the fabric of everyday life. Bureaucrats are everywhere damned; politicians run against Washington; and even popular journalism chronicles the rise of intellectual "neoconservatives" who warn of excessive reliance on the planned amelioration of life's vicissitudes. Whence this paradoxical love–hate relationship to the idea of public policy?

I think the reason is that there has been something irrational about how policy discourse is being conducted in liberal democratic societies. The root of this irrationality is a failure to recognize certain distinctions that provide a conceptual basis for helping to set constitutional limits to public policy enterprises.

Consider these distinctions by way of some questions that should always be asked regarding public policies. When these questions are not asked, policy talk will quickly wander into fanciful visions of social engineering. What are these questions?

1. Is it always appropriate to regard public policies as technical responses to "problems" that are open to technical "solutions"?

2. Is it always appropriate to regard the content of public policies as "social" policies addressing factors that affect only the social structural organization of society?
3. Is it always appropriate to regard public policies as synonymous with government initiated and/or financed policies?
4. Is it always appropriate to regard public policies as pragmatic "practical" responses to specific issues without bothering to debate them in light of evolutionary hypotheses regarding the likely directions being taken by the social and cultural order as a whole?

Generally speaking, most policy discourse in which sociologists are being invited to participate is strongly biased toward affirmative answers to all these questions. It seems to me that as long as this bias continues, discourse about public policy will continue to generate the paradoxical reactions described earlier. The reason is that the public is being conditioned to demand from the government alleviation of its sufferings in technocratic forms that are illegitimate from the standpoint of the public's traditional understanding of freedom in a liberal, pluralistic, and democratic society like the United States. I intend to argue that the bias at issue here is threatening to drive the "public" out of the arena of public policy discourse.

My purpose here will be to respond to the implicit question: what would it be like not to have this bias when discussing policy issues in education (and, by implication, any policy issues)? The answer to this question will be divided into each of the four questions stated earlier, since they form a useful divisional order for our reflections. In the course of discussion, we shall see that these questions imply important distinctions between types of public policies. I hope to persuade the reader that public educational policies promulgated without these distinctions in mind stand an excellent chance of exceeding proper constitutional limits of state interest in the free life of the public. Inchoate awareness of this technocratic threat, I suspect, contributes to the public's ambivalence toward the state's interest in alleviating human suffering and rectifying social injustice.

TECHNICAL VERSUS SENSITIZING POLICIES

Is it always appropriate to regard public policies as technical responses to "problems" that are open to technical "solutions"? Technical policies are activities regarded by their designers as technologically valid means for achieving a finite, specifiable end (the "solution" of a "problem"). Sensitizing policies are activities intended to refine people's awareness that there are some problems that cannot (or should not) be specifiable as having solutions fully attainable by technologically operational means. The elimination of malaria is a goal subject to technical policy in a way that the achievement of the greatest happiness of the greatest number is not. With respect to education, the concept of literacy provides a good example for our reflections on this distinction.

Consider again the 1962 UNESCO definition of literacy:

> A person is literate when he has acquired the *essential knowledge and skills* which enable him to engage in all those activities in which literacy is required for *effective functioning* in his group and community and whose attainments in reading, writing and arithmetic make it possible for him to continue to use these skills towards his own and the community's *development* and for *active participation* in the life of his country. In quantitative terms the standard of attainment in functional literacy may be equated to the skills of reading, writing and arithmetic achieved after a set number of years of primary or elementary schooling (quoted in Curle, 1964:12; emphasis added).

There are many who are tempted to take the quantitative criterion mentioned in the latter part of this paragraph as the basis for a technical definition of literacy. But from the sensitizing point of view, a person can never be definitively regarded as literate, only as more or less so. It can never be objectively clear at any given moment what "essential" knowledge and skills are, nor what "effective" functioning in the "development" of the community means. Furthermore, since what we mean by these things varies with the focus of our moral attention, what we mean by literacy will vary accordingly. This emphasis on the unstable, dynamic, and imperfect character of literacy definitions may seem, and really is, a truism. But it is precisely the drive to express literacy operationally in quantitative terms (skill tests and school levels achieved) that obscures some of the most perplexing problems that a polity faces when it seeks to decide what "minimally literate" means. Making people literate should include bringing them to understand what these problems are and why they are often perplexing and not conclusively answerable in neat technical terms.

SOCIAL VERSUS CULTURAL POLICIES

Is it always appropriate to regard the content of public policies as "social" policies addressing factors that affect the social structural organization of society? Social policies are activities intended to affect socially structured arrangements of some sort such as employment rates, bureaucratic behavior, administrative organization, and financial patterns. Cultural policies are actions designed to affect the concepts and symbols that are embodied in social institutions and justify social practices.[23]

In American society, the formation of systematic, domestic, cultural policy is generally considered an ideologically illegitimate activity for governments, certainly in any form that approaches the tasks of an official propaganda ministry. Yet there are potent examples of cultural policies emanating from both governmental and "private" sources that profoundly affect society's whole symbolic order. Two prime examples are commercial advertising and television programming. These influences are so potent that state action is sometimes demanded to control them. "Truth in advertising" regulations, Federal Communications standards, and the like are efforts to regulate, in the name of public benefits, cultural policies originating in the "private" domain. Despite conceptual difficulties, the effort to distinguish between social policies and cultural policies is justifiable for many reasons. One reason is that the distinction helps us to understand why many social policies are unsuccessful in fulfilling the goals that inspired them. What many people think are social problems requiring social policies to solve them may in reality be cultural problems requiring normative changes in popular values and motives that cannot, in a liberal democracy, easily become a focus for official policies.

With reference to education, the following are some examples of social policies: policies intended to affect the social organization of schooling in terms of administrative structure, teacher roles, and patterns of social integration and segregation; policies that operationally specify the concept of equal opportunity (e.g., taxation and finance patterns, remedial programs, and suburban–urban school district consolidation); policies of certification within the school system and between schools and society; and policies of attendance (compulsory education laws, truancy rules, and exceptions to these regulations).

What, then, are cultural policies? Cultural policies have to do with the symbolic nature and moral ends of education as such. Let us examine some cultural policy problems in education. These are not problems open to strictly technical resolution. They are policy issues nonetheless, and reflection about them serves to sensitize us to the subtleties of public education and its missions in a democratic polity. By way of example, I shall now briefly discuss six questions of cultural policy regarding the proper mission of public education in a liberal democracy.

Should we definitively discriminate education from moral socialization? Socialization is a universal cultural phenomenon. Education, as a more formal institution, is not. Education in the sense of bureaucratically specialized schooling systems is even less common. In the public pursuit of educative ends, what are the constraints on policy that derive specifically from the various possible ways in which the separation of education and socialization evolve in a given society? For example, a democratic society whose state authority is modified by cultural pluralism cannot effectively use a specialized public educational system for nationwide moral socialization much beyond the level of platitudes. To do so would violate that society's consensus on the relative prerogatives of private as against public domains of authority. On the other hand, modern social research has increasingly demonstrated that the schools do in fact socialize children to cultural values in a variety of hidden ways, probably unavoidably so. Do we go on pretending this is not the case or do we face directly the issue of just what universal moral messages we want our public schools to mediate?

Should we educate for action? There is a universally relevant distinction between an educative context as action and an educative context as thought. The contemporary technocratic emphasis on "theory" as something to be translated into "practice" obscures a fact that was more clearly appreciated in premodern societies than in our own. There are certain vitally important things that cannot be taught on the level of theory. This fact leads to the universally relevant educational policy question of how much any given society should emphasize the transmission of literacy through educative action as against didactic verbalization. This question transcends abstract platitudes about learning by doing. It touches on profound matters like the educative significance of adventure, of rituals, of rites of passage, and on the problems of translating into universal educative language the "curriculum" of life's vicissitudes.

Should we educate the emotions? Another question whose significance was better understood in premodern societies than in our own grows out of the universal importance of the double nature of all persons—the intellectual and the emotional dimensions of being human. Accordingly, it may be asked: to what extent should a democratic society that prides itself on rationality institute an educative program for the comprehension of the emotions that is in some ways analogous to logic as the educative language of the intellect? The thorniness of this question should not obscure its importance. In lieu of serious consideration of this question we will continue to hear mindless or outright totalitarian proposals. In recent years, for example, it has been seriously suggested that political candidates be psychoanalyzed, that drugs be used to domesticate aggressive political elites, and that whole populations be subjected to operant conditioning. The outlandishness of such proposals should not blind us, however, to the possible benefits of a more rational effort to focus educative attention upon human emotions.[24]

Should we regulate the balance between oral and visual communication? A scholarly controversy of potential educational policy significance has emerged in recent years out of efforts to probe the different effects of oral as against vision oriented modes of communication. This general topic touches on a number of specific policy issues such as the following. Is there anything more "vicarious" about experience mediated by print as against more direct forms of visional contact? Even if so, is there anything educatively inferior about vicarious experience as against direct experience?[25] What are the effects upon education of the political economy of print, both in the sense of the commercialization of language through the institution of mass textbooks[26] and the possible influences of electronic technology on how print communication is made available for experience?[27] One could go on, but the subject is vast. Perhaps enough has been said to suggest the tasks at hand for educational policy thinkers that are implied by this debate over the social, cultural, and psychic consequences of the historic transitions from oral to written and to film communication.

Should we encourage sophisticated standards of social criticism? It is a universal truism that all educative contexts are embedded in larger institutional environments of social regulation and control. One question this raises for education in a democratic society is the extent to which educative practices should be uncritically subordinate to the central values of these larger environments. With regard to our democratic capitalist society, let us consider two examples of the influence of a capitalist market economy upon education. One, already noted, is the translation of knowledge into a commodity via the mass-marketed "textbook." The other example is the influence of the capitalist division of labor upon the notion of "vocation." It is easy to show that the mass-marketed textbook operates seriously to distort the meaning of ideas. It is likewise easy to demonstrate that the "job" mentality corrupts the integrity of some historically alternative notions of what vocation is all about. The question is: to what extent should educative practices be designed to counteract (and hence conceivably subvert) the dominant values of the larger environment? This is no different from asking what is the proper relationship between education and social criticism. Put still another way, should education have something to do with inducing sensitivity to subordinate or alternative norms and values as well as socially dominant ones? If so, along what range of variation?[28]

Should we educate the mythologic imagination? To pursue all these questions is inexorably to move toward the greatest, most inclusive educational question that a society can confront. This question is the educational status of a society's most sacred myths.

One of the prime features of modern scientific secularism is the fragmentation of the world into two alleged realms: the factual and the moral. Yet scholarly attention to these matters has led to consensus on the enduring importance for societies and persons of what are commonly called "mythological" orientations. Myths are perspectives on reality that, although outside the strictly factual, are nonetheless vitally important influences upon a population's sense of what is meant by reality, truth, time, worth, and so on. The question then becomes: to what extent is it either desirable or possible to attempt a total separation between the factual and the moral as regards mythological themes in the educative processes of a given society?

This question transcends in complexity most of its current doctrinal formulations such as the separation of church and state. By way of example, let us look at three mythic words reflective of modern Western civilization's deepest efforts to represent its destiny in language. These words are "authority," "liberation," and "progress." The problems of liberation and authority, as everyone knows, have received important sociological formulation in the writings of Karl Marx and Emile Durkheim, respectively, specifically in their discussions of alienation and anomie.[29] Anomie is Durkheim's socio-moral diagnosis of a society characterized by a population prone to unlimited aspirations. Alienation, for Marx, is a diagnosis of a society dominated by mystifications that conceal from a population the facts of its own agency in the creation of the social world. A voluminous literature of ideology inspired by these formulations has emerged in recent generations. Should a society seek to make such key formulations part of its program of moral education?

This question applies likewise to the problem of the despair that inevitably attends crises of confidence surrounding a society's eschatological myths. In the West, this is now an issue because of a declining faith in the definability of "progress." Should awareness of sacred and secular eschatologies—their relations to the dynamics of hope and despair, and their connections with the mass movements of yesterday and today—be an important goal of a society's educative efforts? If we do not make this a theme of rational moral education, will the price we pay henceforth be perennial conflict between an official myth of "technology" and recurrent countercultural myths of mindless "revolution"?

I submit that it is foolhardy to search for definitive "technical" solutions to questions such as these. Presuming affirmative answers to all these questions, total official implementation of such policies would take bizarre technicist, if not totalitarian, forms. Presuming a negative answer to all these questions, total official neglect of such issues would undermine the credibility of public education and banish rational discourse about life's most vital matters from the public educative realm. The function of profound cultural policy questions, I suggest, is to stimulate rational public reflection and discourse on the moral order and its transmission to the young. In a democracy, that achievement is part of the answer to any serious question.

PUBLIC VERSUS STATE POLICIES

Is it always appropriate to regard public policies as synonymous with government initiated and/or financed policies? The point of this question is not merely to ask where a policy factually originates, but rather where it should originate according to the political logic of the polity in question. When the people are sovereign in any seriously democratic sense, there are some policies that should not originate in governmental action (e.g., policies regarding the definition and enforcement of religious truth, censorship policies etc.). Further, some perhaps desirable policies could well be subversive of established state interests. A polity which forgets this is in danger of tyranny.

Although these points seem self-evident, they are often ignored in policy discourse. Sometimes this is simply because the state is a definable social institution in a way the public is not, and the state has focusable resources of power and money that the public has not. But the distinction between state interest and public interest is also widely ignored in part because the public is so difficult a concept to define at all, especially in a complex mass society. Yet, despite the belief to the contrary of occasional presidents, the public, not the state, is legally sovereign in American democracy. Historic exhaustion with the conceptual complexities of "sovereignty" is perhaps an explanation but surely not an excuse for the confounding of state and polity. In a liberal democracy the polity and its traditions is the legitimate source of the state, the latter's powers having been regarded as delegated by the former.

The problem, of course, has become: how do we operationally specify the public will if not through the policies of democratically elected governmental representatives? The intractability of this question should not seduce us into permitting the conflation of state and public in face of all the evidence that the state, its own propaganda notwithstanding, is but one more mélange of bureaucratized interest groups. There are some courses of conduct that are neither state policies, nor demands of "private" (parochial) interest groups, but are genuinely public commitments to certain common goods. These commitments are the stuff of traditions, of symbols, of the moral nuances present in the very uses of language. Some of the more noble goals of education are among these, whether these goals be pursued through state controlled or private educational organizations. Is it not these common ends of education, not the source of control, that constitute the true meaning of public education in America? To some extent, this distinction between state and public policies seems to be the real issue often submerged under the political invective of both "liberals" and "conservatives." The issue needs rescue from these exhausted ideological categories.

THE "PRAGMATIST" AND THE "THEORIST": PERSPECTIVES ON THE CONCEPT OF POLICY

Is it always appropriate to regard public policies as pragmatic "practical" responses to specific issues without bothering to debate them in light of evolutionary hypotheses regarding the likely directions being taken by the social and cultural order as a whole? The pragmatist, impatient with theories and muddling through the political process, takes what he can get by way of achieved policies. To the theorist, the result is a patchwork quilt, promising much and delivering little, while exhausting the public's will to believe in rational progress. I may as well admit my predeliction for the theorist's viewpoint. But there is ample ground for sympathy with the pragmatist's suspicion of many ideas that pass in review before him as theories.

There are theorists with a desire for policy "relevance" sufficiently strong to overcome the critical detachment and sense of irony that should be the hallmark of the historically sophisticated theorist.

Since the decline of the Aristotelian perspective on natural law, two major orientations toward what social theory is and can promise have competed for dominance in the West. One orientation has been the positivist tradition (in its nineteenth century Saint-Simonist sense). In this view, the task of theory is to discover the laws of social order and of social development and to design policy according to their logic. This attitude was present in classical sociology as late as the early works of Emile Durkheim, and as we saw, is present today in much of contemporary systems theory. The classically definitive break with this prescriptive form of positivism occurs in the tragic perspective of Max Weber's sociology. Here the observer becomes an ironic commentator on history as the subversion of abstract ideals and personal commitments. To the ironic theorist the prescriptive positivist becomes the technocrat at best and the totalitarian engineer of conscience at worst. To the pragmatist, the ironic theorist becomes a passive and elitist witness to avoidable mass suffering, the aesthete connoisseur of evil.

I know of no way to resolve this classic tension. But let us "do" a little theory with respect to a baffling moral problem afflicting the whole enterprise of public policy: how do we operationally specify what is meant by minimal legitimate state interest in public education in the context of American liberal democracy? This too can hardly be answered definitively here. But what I do wish to demonstrate is the claim that the question cannot even be asked without reference to theories about what the social and cultural orders are and in what directions, so to speak, they are evolving. Such questions are difficult and press the patience of pragmatists whose temporal sense is constrained by the dynamics and possibilities of a given political administration. But not to pose the larger theoretical problems is to move forever on the surfaces of historical immediacy, bartering away for small gains many elements of aspiration and tradition whose rhythms are only detectable from a larger perspective. I intend to pursue now the question of how to define the present minimally legitimate state interest in American public education.

CURRENT STATE INTEREST IN AMERICAN PUBLIC EDUCATION. The situation with respect to state interest in American public education is perhaps best exemplified by an important Supreme Court decision in the case of *Wisconsin v. Yoder*.[30] Members of the Amish religion had been convicted of violating Wisconsin's compulsory school attendance law. The Supreme Court decided in favor of the Amish respondents' claim that application of the compulsory school attendance law violated their rights under the free exercise clause of the First Amendment.

Several arguments in the majority opinion deserve notice as regards our present discussion because they represent a reasonable indicator of American society's "official" position on some of these matters.

> The unchallenged testimony of acknowledged experts in education and religious history, almost 300 years of consistent practice, and strong evidence of a sustained faith pervading and regulating respondents' entire mode of life support the claim that enforcement of the State's requirement of compulsory formal education after the eighth grade would gravely endanger if not destroy the free exercise of respondents' religious beliefs (Wisconsin v. Yoder:219).

We have here the Court's effort to define a "world" whose claims could contravene usual state interests expressed in compulsory education laws. The definition is quite stringent, resting on a rather strict sense of the term religion:

> The very concept of ordered liberty precludes allowing every person to make his own standards on matters of conduct in which society as a whole has important interests. Thus, if the Amish asserted their claims because of their subjective evaluation and rejection of the contemporary secular values accepted by the majority, much as Thoreau rejected the social values of his time and isolated himself at Walden Pond, their claims would not rest on a religious basis. Thoreau's choice was philosophical and personal rather than religious, and such belief does not rise to the demands of the Religion Clauses (ibid.:215–216).

What, according to the Court, is the state's interest in education?

> The State advances two primary arguments in support of its system of compulsory education. It notes, as Thomas Jefferson pointed out early in our history, that some degree of education is necessary to prepare citizens to participate effectively and intelligently in our open political system if we are to preserve freedom and independence. Further, education prepares individuals to be self-reliant and self-sufficient participants in society. We accept these propositions (Wisconsin v. Yoder:221).

This almost unimpeachable abstract statement of aims for education in a democratic society is given concrete limits in this case worthy of note.

> The State attacks respondents' position as one fostering "ignorance" from which the child must be protected by the State. No one can question the State's duty to protect children from ignorance but this argument does not square with the facts disclosed in the record. Whatever their idiosyncracies as seen by the majority, this record strongly shows that the Amish community has been a highly successful social unit within our society, even if apart from the conventional "mainstream." Its members are productive and very law-abiding members of society; they reject public welfare in any of its usual modern forms (ibid.:222).

And again:

> There is nothing in this record to suggest that the Amish qualities of reliability, self-reliance, and dedication to work would fail to find ready markets in today's

society. Absent some contrary evidence supporting the State's position, we are unwilling to assume that persons possessing such valuable vocational skills and habits are doomed to become burdens on society should they determine to leave the Amish faith, nor is there any basis in the record to warrant a finding that an additional one or two years of formal school education beyond the eighth grade would serve to eliminate any such problem that might exist (ibid.:224–225).

What we see here is an official view on the nature of the ignorance against which state interest must intervene. It is a conception of ignorance that is limited to inability to take part in the market economy. Apart from membership in an alternative religious world, the unlikelihood of being a "burden on society" is the sole definition advanced by the Court of "self-reliant and self-sufficient participants in society" (Wisconsin v. Yoder:221). In fact, in this era of late industrial technological society, the Court seemed obliquely to endorse Thomas Jefferson's agrarian based philosophy of education by equating the Amish with Jefferson's "sturdy yeomen" (ibid.: 225–226).[31]

Now, in noting all this, I do not mean to argue that the Court's decision was improper. It was quite in line both with democratic respect for cultural pluralism and with a liberal society's minimalist notion of state interests.[32] But I do wish to point to four implications of the case for those interested in drawing distinctions between public as against merely state educational policy.

1. Although a liberal capitalist democracy can define the state's interest in education as based on a desire "to prepare citizens to participate effectively and intelligently in our open political system," what this means in practice will turn out to be conservative and limited. This is because the state's interest will represent the social status quo, not some vision of what society *could* be if its citizens were educated to participate in the fulfillment of civilization's visions of progress rather than merely society's factual organization.

2. It is difficult to see how the rhetoric of freedom in a capitalist democracy's educational policy can officially move beyond how the Court has conceived it in this case. Freedom, in *Yoder*, turns out to be the right of plural "worlds" to coexist under state tolerance so long as they are in some sense genuine worlds and do not produce individuals who are social welfare burdens. Whenever this can be demonstrated, as it was for the Amish, it becomes "incumbent on the State to show with more particularity how its admittedly strong interest in compulsory education would be adversely affected by granting an exemption" (Wisconsin v. Yoder:236). It is unlikely that the state could make such a showing without a more positive conception of society than a population adapting to the facts of the market economy. The state would have to represent a society organized to harness science and technology for standards of personal freedom and development more transcendent than economic survival and material aggrandizement.

3. It is difficult to see how the rhetoric of equality in a capitalist democracy's educational policy can officially move beyond the notion of equal access to two goals implicit in the Court's decision. One goal is the right to socialize one's children into plural worlds, achieved by denying the state the right officially to promulgate religious doctrines in state schools. The other goal is adequate preparation for rational occupancy of a position in society's labor market. For any broader conception of educational equality (such as the right of equal access to the most elite available standards of preparation for personal development and cultivation), there would have to be a guiding vision beyond the status quo.

4. Given what we know about the effects upon society of science, technology, and social stratification, most worlds like the Amish are doomed to existence outside secular history. The sort of cultural pluralism represented by the Court's decision actually involves a trade-off. It is an incipient form of the trade-off described in my depiction of the libertarian technicist society in Chapter 2 of this book. Why is this so? It is not in the state's official interest to educate persons for a standard of citizen participation beyond adaptation to the status quo. Therefore, worlds that reject modern secular civilization have the right to unmodified existence even though they produce people unqualified to take part in civilization and its aspirations. To the extent that present pluralist and countercultural tendencies continue, we will witness the revival and/or emergence of myriad enclave worlds. Their partisans will "drop out" of history's complexities, leaving—as the price of tolerance—the management of society to state systems engineers. In time, the libertarian technicist model becomes accepted political philosophy.

The significance of these implications is this: it is not reasonable to regard as part of the state's interest in education the development of citizens profoundly enough educated to be competent participants in a *non*-technicist environment. If *Yoder* is paradigmatic of how far the state interest goes in a liberal democracy, then state interest in education is compatible with a pluralist, and to some extent, libertarian conception of society. This is in line with the minimalist idea of the state. Since, as we have seen, there are societal problems that people are willing to concede require state intervention, such interventions are likely to be justified on utilitarian (technological) grounds. The libertarian technicist scenario is the only one that preserves the balance between an ideology of libertarianism and state intervention based on meliorist or utilitarian grounds.

A genuine antitechnicist educational program would, in contrast, presuppose a social philosophic vision stressing common appropriation of all technologies for the sake of some collective drama of rational progress. A political philosophy reflecting this sort of vision would entail a much more positive notion of state interests than the minimalist liberal conception allows for. Nothing is gained by sentimentally closing our eyes to the fact that both technicist and antitechnicist visions of society require an actively interventionist state. Technicism implies an interventionist state as the mediator of techniques and expertise. An antitechnicist vision of society implies an interventionist state both as long-range "planner for freedom"[33] and as redemptive "liberator" of persons from technicist tendencies in society itself. The problem here would be the classic one of having to force men to be free. The only surprise-free resolution of these issues I can envision is the evolution of the libertarian technicist society modeled in Chapter 2.

Conclusion

In this chapter I have advocated a view of public educational policy discourse that is closer to scientifically informed political philosophy than to science as such.[34] By implication, I have argued several virtues of this approach. It is a prophylactic against premature public reliance on technocratic expertise. It helps to explain popular ambivalence toward the concept of public educational policy. This ambivalence I have ascribed to a systematic bias among experts and laypersons alike toward viewing public policy as comprising primarily government initiated, technological, social organization oriented, and pragmatic (crisis responsive) activities. In place of this bias I have urged that we regard public policy discourse as an invitation for intellectual activity more fundamental than technological problem solving. It could be conversation about the nature of our shared culture, not just game planning for organizational tampering. It could be democratic political conversation, not just proposal writing for governmental funding. And finally, it could be a less pragmatic and more historically self-conscious, perhaps even prophetic, mode of reflection. In this context, perhaps, it would not be far-fetched to hope that pragmatists and theorists might learn to address each other with more respect. If theorists would presume less about the technocratic translatability of their ideas and would apply to pragmatist activity the criteria of prudential wisdom, then perhaps the pragmatists would show greater receptivity to genuine intellectual wisdom on the part of thinkers.

Notes

1. How complex and controversial a concept indoctrination is can be gleaned from a perusal of the essays in the volume edited by Snook (1972).

2. The loss of awareness among educators of all sorts about the historicity of the very idea of curriculum has led to an impoverishment of discourse about the role and the limits of the idea of curriculum in our culture. Every educational thinker

would benefit greatly from reading and contemplating the implications for our time of Ong's great study of the history and sociology of Ramism (1974).

3. In recent years there has been a growing literature of what might be called the "hidden agenda" aspects of schooling, much of it produced by Marxists. One of the non-Marxist examples is the well-known book by Dreeben (1968), which is grounded in the structural-functional sociology of Talcott Parsons.

4. An excellently detailed account of why this approach is indeed so irrational can be found in Hirst's (1972) critique of the 1946 Harvard Committee Report entitled "General Education in a Free Society."

5. Philosophy of education is sometimes written as though education is to prepare people for a timeless, stable world. Sociology of education is sometimes unduly influenced by those who think that every new fashion signals a world-historical environmental change to which education must necessarily respond. The four contexts I am to discuss seem to provide reasonable reference points for the study of both stability and change. By studying the interaction of events across these four contexts, we can isolate some objective characteristics of the factual and moral contradictions that beset the complex environments of which education is a part. Philosophy of education can be thought of as the discipline that inquires into criteria about what it means to educate persons for effective agency in these diverse contexts. Sociology of education inquires into the social conditions that generate, alter, or destroy these contexts and the factual implications of all this for education. The reader is reminded that although my definitions are within reasonable range of traditional sociological usage, the use I make of these concepts here is my own. It would be beside the point, I think, to oppose these arguments on the ground that there are other possible definitions of these terms than the ones given here. I am also aware that these contexts are not separate worlds and that education cannot really be as clearly separated from its cognates as has been done here. One must keep in mind the point of this analytical exercise: as much as possible I am trying to set the stage for conceptually reserving to education the mission of immunizing the psyche against technicization. For this reason it seem to me desirable to make these distinctions. I am trying to clarify the ancient intuition that one may be socialized, trained, indoctrinated, and schooled without necessarily being educated.

6. Naturally, many of the classical ideals of *paideia* come to mind here (Jaeger, 1942–1944), as do the ideals implicit in Cicero's conception of the orator.

7. As the reader will have guessed, the sense here is of society as Toennies's notion of *Gesellschaft*. The thinker who more than any other wrestled with education for such a context, both directly in his work on moral education (1973) and in virtually all his works indirectly, was Durkheim.

8. Naturally this has taken different forms, depending upon the beliefs prevalent in different cultures. But such a vision is as much present in the more universalistic, and eventually Stoic, meanings of the classical idea of *paideia* as it is 2,000 years later in Hegel's philosophy of education.

9. The exemplary analyst of this sense of contemporary crisis was Hannah Arendt. Her greatness is partly reflected in the fact that she was one of the very few writers using terms like "authority," "crisis," and "tradition" whose work yet defies simplistic ideological classification as "conservative."

10. The UNESCO definition of 1962 and the approach behind it has been subjected to strong criticism even within the ambience of existing notions of development, for example, by Curle (1964). One of the more insightful critiques of standard thinking on educational planning is a study of OECD planning rhetoric by Webster (1971).

11. To my mind, a classic example of exactly these "sins" committed in the grossest manner is Bronfenbrenner's popular work on Russian and American education (1970). In its total confounding of the concerns of education with those of socialization—a confounding baptized by all the latest terminology of social psychology—the book is an exemplar in educational technicism as defined here. The failure even to mention, in a book half devoted to Russian education, that the Soviet Union is a totalitarian society is as much a mark of American political innocence as the earlier tendency among some other "humanists" to regard Soviet communism as the wave of the future. Bronfenbrenner's book was, ironically enough, published in the same year that Gouldner published a book in which these apposite remarks appear (1970).

> The meaning of a rapprochement between Soviet and American sociology, or between Academic Sociology and Marxism, will depend greatly on the extent to which each is committed to an administrative or managerial sociology. An administrative sociology is essentially an instrument for making the status-quo work better . . . in short, an administrative sociology misses the nature of the competition and struggle between alternative solutions (ibid.:475).

12. Still relevant in this regard is Aristotle's discussion of the significance of these discontinuities in Book VI of his *Nicomachean Ethics*. For a poignant contemporary reflection on the problem of whether knowing has much to do with doing as regards the avoidance of doing evil see Arendt (1971).

13. One of the leading works of scholarship ever done with regard to this program is Ong's study of the cultural evolution of Ramism in Western education (1974). In various publications he has spelled out Ramism's significance for our understanding of the contemporary technicist nature of education.

14. Of notable significance here is the volume of studies on "thinking" in Western and non-Western societies edited by Horton and Finnegan (1973).

15. With respect to the moral significance of sociolinguistics as seen by one of its creators see Hymes (1973). See, too, Habermas's effort to conceptualize the meaning of communicative competence as a stage in the evolution of ethics and reason (1970).

16. A bibliography and critique of this research can be found in Karabel and Halsey (1976).

17. For a review of the variety of activities that are hermeneutical in nature, and an effort to define an underlying methodological unity among them, see Palmer (1969). For the notion of studying social action as a text analogue, a matter of great importance to sociologists, see Ricoeur (1971). There is a growing literature of what could be called a hermeneutical sociology being published in England, largely by Routledge & Kegan Paul.

18. Sociologists of knowledge could have some interesting things to say about the implied contrast here between the social origins of efforts in literacy training. The "medical" approach is appropriate for literacy specialists who come out of the

mission field. Until very recently, missionaries saw themselves as bringing enlightenment, civilization, and salvation to those without them. Freire's orientation, on the contrary, is appropriate to the social outlook of rising ex-colonials developing a revolutionary consciousness.

19. In this connection, see Freire's study of the concept and semantics of "extension" education (1973:93–164).

20. This orientation of "waiting" is perhaps most marked in Heidegger's thought. It was apparently sustained through the end of his life as evidenced in his final interview (1976).

21. While raising these warnings about elitism, however, I want to dissociate myself from the totally unsympathetic attack on Freire by Berger (1974:111–132). I regard Berger's attack as lacking any appreciation of the nihilistic relativism necessary for his own complete rejection of the concept of false consciousness. (In this matter I concur completely with the critique of Berger's thought by Carveth, 1977.) Even Berger, it seems, has difficulty maintaining his relativism since he allows an important exception to his "postulate of the equality of all empirically available worlds of consciousness" (1974:116).That exception takes the conceptual form of "information": "If the hierarchical view of consciousness simply referred to levels of information of specific topics, there would be no need to quarrel with it" (ibid.:114). How is it possible to argue the superiority of information (especially, as he does, on "topics" like "economics") while remaining totally innocent of any belief in some kind of "cognitive hierarchy"? And *is* it really the case, as Berger claims, that peasants have "far superior information and perspectives" on "topics" like botany, soil, weather, and the intricacies of kinship, not to mention the "true significance of dreams" (ibid.:114–115)? In accusing Freire of paternalism, elitism, and cognitive imperialism, Berger ignores Freire's life. Having himself survived poverty and oppression, Freire does not assume that the word "equality" is necessarily incompatible with recognizing the existence of impoverished and oppressed forms of consciousness.

22. Science fiction writers have begun treating "new man" literacy notions. An especially interesting example is Gunn's *The Joy Makers* (1961).

23. The point of this distinction is the possibility that the content of some policies is cultural, not social. Some thinkers regard a distinction between the "social" and the "cultural" as trivial or unoperational. The position one takes on this conceptual issue has implications for one's sense of what policy means, and what the objects of policy are. I regard the distinction as very important. Of course, cultural analysis of policy issues is not exactly a novel thought. Numerous thinkers, especially those variously influenced by Marxist critical humanism, have criticized the social policies of the Welfare State. It should perhaps be stated at the outset that my own focus on the need for cultural analysis does not stem from the same ideological premises. I do not, of course, oppose emancipatory social criticism; rather, the theme of emancipation is for me not at the ideological center stage. I persist in believing that the republican traditions of the American constitutional order can assimilate most people's understanding of the good society, including mine. As for many, however, this belief is tinged with ambivalence. Hence I find it distressing that the effort to examine these traditions to see if this is so seems to me to have been largely abandoned by both "conservatives" and "liberals" alike. Part of the value commitment of any sort of traditionalist is the belief that people exist whose sense of

identity and normalcy, however inchoate, remains informed by some shared sense of communal fate. Such people are classically referred to as a "public." For a public, traditions can serve as a stable reference point both for conservation and for criticism. A public can forget itself and be recalled to its identity; it can be subverted by false counsel; it can be disordered and transmuted into masses. Whatever its vicissitudes, a public is first of all a cultural entity: it knows itself in consciousness, its expressions are symbolic. Its interests represent the most universal themes of a tradition, defining the individual as a member of something more permanent and more precious than the momentary organizational dynamics that determine his daily fate. The latter is the realm of power, but the former is where we search for authority. In that spirit of culture this essay is written.

24. The position with which I find myself in agreement on this matter is set forth in Peter's essay on "The Education of the Emotions" (1972). Also of direct relevance to such a project is Solomon's philosophical exploration of *The Passions* (1976).

25. Aside from the well-known works of Marshall McCluhan and his intellectual precursor, Harold Innis, see Ong (1969), Havelock (1963), and some of the essays in the collections edited by Goody (1968), Horton and Finnegan (1973), and Disch (1973).

26. Many of the points made about the political economy of language from a contemporary Marxist point of view by Marcuse (1964) are placed in a much broader framework by Ong in his cultural-historical studies of the rise of Ramism (1974).

27. Unfortunately, this debate between the partisans and critics of print is being carried on at a very incomplete scope. What is left out of account by those who are attacking or defending print in the name of direct and indirect experience is the larger sociology of abstraction. This argument has been made by both Marxist and non-Marxist sociologists of culture, with respect to everyday life in contemporary industrial societies. Aside from the obviously relevant contributions of classical sociologists like Durkheim and Weber see Zijdereld (1970) and Lefebvre (1971) for a non-Marxist and a Marxist treatment, respectively, of the abstract nature of "direct" experience in contemporary societies.

28. It has been somewhat hesitantly debated, from time to time, whether propaganda detection, both commercial and political, should be overtly taught in the schools. If undertaken seriously, such an enterprise would clearly have subversive effects upon much of the dominant social structure. It is not surprising that such efforts are not a central part of the curriculum.

29. The obscuring of the original significance of the concepts of anomie and alienation by way of tendencies toward psychological reductionism in American social science is discussed by Horton (1964).

30. *Wisconsin v. Yoder (1972)*. I say this case is a reasonable indicator for our use in this context, not the only possible one. For some of the complex possibilities of how this case can be regarded from the standpoint of state intervention into private affairs, see Baskin (1974) and Knudsen (1974). While these commentaries raise qualifications, nuances, and perhaps other possible interpretations of the interests of the particular parties involved in this case, I do not see anything in the commentaries that directly contradicts the analysis that follows here.

31. The main basis for Justice Douglas's partial dissenting opinion was the fact that some of the children's rights to choose for themselves had been overlooked.

32. I am taking for granted here the central importance for liberal political ideology of the minimalist notion of the state. Current American polemic between "conservatives" and "liberals" does not strike me, in contrast with the great European political-philosophic conflicts of the past, as anything but a quarrel between different conceptions of the minimalist mission of the state. Conservatives stand for a more pure nineteenth-century laissez-faire liberalism than do contemporary liberals who would use the state for meliorist efforts to protect people against the turmoil of market forces. It is difficult to imagine any major American political party, much less the judicial system, departing from this tradition to argue for some transformative mission for the state.

33. I mean by this phrase the sense intended by Mannheim in his work on social planning in democratic societies (1949).

34. For an example and relevant bibliography of what a more purely scientific concern for the meaning of "public policy theory" is, see Greenberg (1977).

Epilogue

TOWARD A POLITICS OF COMPETENCE

This book is about a form of mystification. If one had to state the essence of mystification briefly, a reasonable definition would be that mystification is concealment of the processes whereby things happen or are made to happen by means of intended actions. Concealment should not be regarded as necessarily a matter of deliberate design. Often it is an inadvertent result of the way environments (including linguistic environments) evolve or diffuse from one setting to another. Intentional or not, however, mystification must be met with social criticism.

Antitechnicist social criticism should have as its aim to reverse the current tendency toward "disarming" the citizenry of linguistic, cognitive, and social competences. A program to preserve and increase the competence (and therefore the dignity) of the person as a political agent can, in my judgment, be carried to considerable lengths within the present social order of the industrial democracies. Many of the ways in which giant social institutions are disarming the citizenry of their competences are questionable from the standpoint of the cultural and juridical traditions of these societies. From this standpoint (which assumes the ideological sincerity of these "camps"), many contemporary "conservatives," "liberals," and "radicals" have more in common than their mutual hostility allows them to consider. Be that as it may, let us conclude this book with some very brief comments on possible countertechnicist strategies of demystification.

The movement to criticize and demystify real technological devices is gaining momentum among critics and designers. To some extent, cybernetic systems analysis appears congenial to proponents of this movement who believe that cybernetic design will facilitate radical technological, and there-

250

fore socio-political, decentralization. Their vision is of a return to institutions on a human scale.[1] The movement toward technological decentralization, with or without cybernation, is congenial to certain American traditions like communalism and anti-urbanism. In its present form, much of this movement seems quite reconcilable with what I have called libertarian technicism. A more precise sociological and political-philosophical imagination will have to be exercised by those who want to preserve the demystification intentions of this movement among technologists and ecological designers without encouraging the romantic pastoralism that seems to go along with it. Since my major focus is not real technology, however, but technicism, I shall not consider this issue further here. This book has largely been about pseudodevices on the level of language. So my subsequent remarks are restricted to the demystification of linguistic technicism. The rational ends of social criticism would be furthered by persistent attention along three fronts: the design of social research methodology; the further institutionalization of linguistic accountability; and the critique of irrational specialization. Much has already been said and implied in this book about technicist and nontechnicist research orientations in social science and education. Therefore, there is no need to belabor this topic. About the other two fronts of criticism, however, a brief word is perhaps in order.

Moral Accountability and Linguistic Criticism

The entire relationship between persons and language needs reexamination in light of the demonstrated power of language to manipulate fantasies, create pseudoevents, and control symbolic experiences of all sorts. A politics of competence begins with the question of whether we can any longer afford to allow agents of powerful institutions to do anything they want with language, the most significant of all human symbolic phenomena. This question is not raised to encourage ventures into censorship unless "truth in advertising" or libel legislation can be said to be censorship.[2]

What is needed, I think, is a logic of moral accounting integrated into our juridical theory of legitimate institutions. It is a striking social and cultural fact that Western civilization, having once had a logic of moral accounting, has totally abandoned the project instead of reforming and updating it. I refer to the medieval tradition of casuistry. The fact is that at least some acts committed by men of power once were subjected to a logic of justification wholly devoted to practical reasoning in the light of certain broad moral principles (and, less desirably, of dogmas). Major examples were debates over the meaning of just price and just war.[3] For various complex reasons, including misuse of the relevant casuistical techniques, this approach to moral accounting was abandoned. It has never been replaced, and today uninformed people virtually equate casuistry with sophistry. Yet there has been in American history at least one important episode of public debate on a level of the most creative casuistical standards: the debate over ratification of the United States Constitution, whose monument is the *Federalist Papers*. Today there are no standards of moral-linguistic accounting. In their place we have mindless appeals to slogans like "national security," "the free world," "law and order," "liberation," and "self-actualization."[4] A politics of competence begins by recognizing these as clichés and proceeds to experiment with their translation into accounting procedures for practical reasoning.

We would do well to take as the first illustrative candidate for such close attention the concept of social "complexity" itself. In the name of this concept many compromises have been imposed upon personal dignity and social progress. One need not go all the way down the anarchist road with radical critics like Murray Bookchin in order to appreciate the vital challenge to our imagination for social design posed by this question:

> Is society so "complex" that an advanced industrial civilization stands in contradiction to a decentralized technology for life? . . . Modern society is incredibly complex, complex even beyond human comprehension, if we grant its premises—property, "production for the sake of production," competition, capital accumulation, exploitation, finance, centralization, coercion, bureaucracy and the domination of man by man. Linked to every one of these premises are the institutions that actualize it—offices, millions of "personnel," forms, immense tons of paper, desks, typewriters, telephones, and, of course, rows upon rows of filing cabinets. As in Kafka's novels, these things are real but strangely dreamlike, indefinable shadows on the social landscape. The economy has a greater reality to it and is easily mastered by the mind and senses, but it, too, is highly intricate—if we grant that buttons must be styled in a thousand different forms, textiles varied endlessly in kind and pattern to create the illusion of innovation and novelty, bathrooms filled to overflowing with a dazzling variety of pharmaceuticals and lotions, and kitchens cluttered with an endless number of imbecile appliances (Bookchin, 1971:136–137).

Critique of Irrational Specialization

The office of citizen requires critical examination from an important sociological perspective: the structure of specialization. The social division of labor is not just one among a number of structural facts about society. In modern societies the social division of labor is one of the primary structural prisms through which language as symbolic experience and action is refracted. We now know some things about the division of labor in industrial societies that help explain how extreme specialization threatens the integral competence of the person as political agent.

For example, specialization can proceed to a degree that is demonstrably irrational in its effects. Social structural specialization is often replicated on the level of subjective consciousness. Beyond a certain point, specialization can induce in a population general incompetence and helplessness rooted in a sense that it is dangerous to act except in professionalized, role defined, and credentialed settings.

The sociological dimension of a politics of competence should therefore be a concerted pressure toward the definition and reversal of irrational specialization in all aspects of life. This is tantamount to a general critique of unlimited professionalism.[5] With respect to language, such a critique encourages a never-ending examination of technical vocabularies and their uses for social control through mystification. Without such political attention to the proper limits of professionalism, most cultural advances in symbolic competence will prove socially and politically fruitless.

It is difficult to see how efforts like these can be mounted in society if they do not begin in the universities. Yet these now produce more specialists and technocrats than genuinely educated people, more "experts" than "thinkers." As a civilization, we are in danger of cultural amnesia, of forgetting things that our best educated ancestors understood well. Many of us float on the ocean of the present, clinging like shipwrecked sailors to the flotsam of great ideas systematically developed by thinkers of the past. Meanwhile, outside the universities in society at large, we all increasingly come to experience language and its moral possibilities in only two basic ways. One is as a reflex of our function in some organization (the "job morality"). The other way is in the form of pieties about "values." Speech appropriate to the job morality is largely that of technicist "solutions" to "problems" whose larger dimensions we barely comprehend. Speech appropriate to values is largely that of abstract virtuous intentions. In neither case do we really have much experience today with speech about action, only that about "functions" and "ideals."

It would be well for us all to begin our third century as a nation by coming to understand the implications of this momentous cultural fact: for most of us, life is now dominated by the social reflections of our destiny as the technicians and functionaries we have largely become. We think of this as "the real world." Yet, if a real world is one in which we choose, act, and accept reasonable consequences, then very few of us live in a real world anymore. To be educated is ultimately to be fit for conversation not merely about abstract intentions but about the contingent social practice of life. Without such conversation, no one can hope to learn what he really wants. And people who know not what they want cannot hope rationally to pursue happiness.

Notes

1. For the connections between cybernation and the problem of human scale see Winthrop (1968), Theobald (1965), and Bookchin (1971). Although he does not mention cybernation and would probably be dubious about the claims of its partisans, one of the better recent books on the connections between technicism and loss of human scale is Illich (1973). Schumacher's book on intermediate technology, largely an attack upon the primacy of economistic thinking and its relation to the loss of human scale, has little to say one way or the other about cybernation (1973).

2. The spirit in which we need worry about this is that which informs much of the content of Steiner's volume of essays on *Language and Silence* (1970).

3. For some examples of casuistical problems and debates in Western history see Nelson (1969), Hanke (1959), and Johnson (1975).

4. In recent years there have been signs among some philosophers and social scientists of renewed interest in the casuistical uses of knowledge and the resources of the intellect. One such indication is the journal *Philosophy and Public Affairs*. Other examples include the volumes published in recent years on "normative" social science, on political obligation, and on modern conditions relative to classical political and philosophical perspectives. There has even been a collection of essays trying to deal with the implications for "conscientious action" of the Pentagon Papers (French, 1974). It is not clear, however, that these efforts are having much effect in public life.

5. Bledstein's attack on "the culture of professionalism," although it goes further than I feel comfortable about, is an important scholarly indicator of a growing legitimacy crisis among intellectuals regarding the professions and their powers in the division of labor of modern societies (1976).

Bibliography

This bibliography is not intended to be exhaustive in any sense. It merely records references mentioned in the text. Over the years of preparation for this book, I have read a list of materials, both directly and indirectly relevant, that is much larger than this bibliography. This includes popular books in the antitechnology and antiscience mood that dominated the 1960s counterculture scene, philosophical works, works on systems theory and systems philosophy, sociological studies bearing on these matters, historical works, and so on. To cite them would be pointless. I am aware, too, that this bibliography does not include important literature on the numerous specialized intellectual problems that I have touched upon in my limited journey through the problem of technicism. Some of this literature is familiar to me, much doubtless not.

ABEL, LIONEL. 1970. "Sartre versus Levi-Strauss." *Claude Levi-Strauss: The Anthropologist as Hero*. Edited by E. N. Hayes and Tanya Hayes. Cambridge, Mass.: M.I.T. Press.

AFANASYEV, V. G. 1971. *The Scientific Management of Society*. Moscow: Progress Publishers.

ALLISON, GRAHAM T. 1971. *Essence of Decision*. Boston: Little, Brown.

APTER, MICHAEL J. 1966. *Cybernetics and Development*. New York: Pergamon.

ARENDT, HANNAH. 1958a. *The Origins of Totalitarianism*. New York: Meridian Books.

_____. 1958b. *The Human Condition*. New York: Doubleday.

_____. 1963a. *Between Past and Future*. New York: Meridian.

_____. 1963b. *On Revolution*. New York: Viking.

_____. 1963c. "Man's Conquest of Space." *American Scholar* 32, no. 4 (Autumn).

_____. 1971. "Thinking and Moral Considerations." *Social Research* 38, no. 3 (Autumn).

ARISTOTLE. 1946. *The Politics*. Translated by Ernest Barker. New York: Oxford.

_____. 1954. *The Nicomachean Ethics*. Translated by David Ross. New York: Oxford.

ASHBY, W. ROSS. 1961. *An Introduction to Cybernetics*. London: Chapman & Hall.

AVINERI, SHLOMO. 1970. *The Social and Political Thought of Karl Marx*. Cambridge, Eng.: Cambridge University Press.

BAKER, MICHAEL. 1975. *Condorcet*. Chicago: University of Chicago Press.

BAKKER, DONALD PAUL. (n.d.). "The Philosophical Discussion of the Nature of Information in the U.S.S.R." Master's thesis, Department of Political Science, Columbia University.

BARNETT, HAROLD J., AND MORSE, CHANDLER. 1963. *Scarcity and Growth*. Baltimore: Johns Hopkins University Press.

BARRACLOUGH, GEOFFREY. 1971. "Hitler's Master Builder." *New York Review of Books*, 15, 12 (January 7).

BARRETT, WILLIAM. 1972. *Time of Need: Forms of Imagination in the 20th Century*. New York: Harper & Row.

BASKIN, STUART J. 1974. "State Intrusion into Family Affairs: Justifications and Limitations." *Stanford Law Review*. 26 (June).

BEER, STAFFORD. 1964. *Cybernetics and Management*. New York: Wiley.

———. 1974. *Designing Freedom*. New York: Wiley.

BENDIX, REINHARD. 1960. *Max Weber: An Intellectual Portrait*. Garden City, N.Y.: Doubleday.

BEREITER, CARL. 1971. "Education and the Pursuit of Reality." *Interchange* 2, no. 1.

BERGER, PETER. 1974. *Pyramids of Sacrifice*. New York: Basic Books.

BERGER, PETER, BERGER, BRIGITTE, AND KELLNER, HANSFRIED. 1973. *The Homeless Mind*. New York: Random House Vintage.

BERNSTEIN, RICHARD J. 1971. *Praxis and Action*. Philadelphia: University of Pennsylvania Press.

BERTALANFFY, LUDWIG VON. 1968. *General System Theory*. New York: Braziller.

BLEDSTEIN, BURTON. 1976. *The Culture of Professionalism*. New York: Norton.

BLUM ALAN F., AND McHUGH, PETER. 1971. "The Social Ascription of Motives." *American Sociological Review* 36, no. 1 (February).

BOGUSLAW, ROBERT. 1965. *The New Utopians*. Englewood Cliffs, N.J.: Prentice-Hall.

BOORSTIN, DANIEL J. 1964. *The Image*. New York: Harper.

BORGMANN, ALBERT. 1972. "Orientation in Technology." *Philosophy Today* 16, no. 2/4 (Summer).

BOURKE, VERNON. 1964. *The Will in Western Thought*. New York: Sheed.

BOWEN, HOWARD R. 1977. *Investment in Learning: The Individual and Social Value of American Higher Education*. San Francisco: Jossey-Bass.

BRENNER, MICHAEL J. 1969. *Technocratic Politics and the Functionalist Theory of European Integration*. Ithaca, N.Y.: Center for International Studies, Cornell University.

BREWER, GARRY D. 1973. *Politicians, Bureaucrats, and the Consultant*. New York: Basic Books.

BROEKMAN, JAN. 1974. *Structuralism: Moscow, Prague, Paris.* Dortrecht: Reidel.

BRONFENBRENNER, URIE. 1970. *Two Worlds of Childhood: U.S. and U.S.S.R.* New York: Russell Sage.

BROWN, RICHARD H. 1977. *A Poetic for Sociology.* Cambridge, Eng.: Cambridge University Press.

BRUNNER, JOHN. 1969. *Stand on Zanzibar.* New York: Ballantine.

BUCKLEY, WALTER. 1967. *Sociology and Modern Systems Theory.* Englewood Cliffs, N.J.: Prentice-Hall.

_____, ed. 1968. *Modern Systems Research for the Behavioral Scientist.* Chicago: Aldine.

BUNGE, MARIO. 1969. "The Metaphysics, Epistemology, and Methodology of Levels." *Hierarchical Structures.* Edited by L. L. Whyte, A. G. Wilson, and D. Wilson. New York: American Elsevier.

CAHN, EDMOND. 1949. *The Sense of Injustice.* Bloomington, Ind.: Indiana University Press.

CARROLL, JOHN B. 1974. "The Potentials and Limitations of Print as a Medium of Instruction." *Media and Symbols.* Edited by David E. Olson. Chicago: University of Chicago Press.

CARVETH, DONALD. 1977. "The Disembodied Dialectic." *Theory and Society* 4, no. 1 (Spring).

CASSIRER, ERNST. 1953. *Substance and Function, and Einstein's Theory of Relativity.* New York: Dover.

_____. 1960. *The Logic of the Humanities.* New Haven, Conn.: Yale University Press.

_____. 1964. *The Individual and the Cosmos in Renaissance Philosophy.* New York: Harper.

CAWS, PETER. 1970. "What Is Structuralism?" *Claude Lévi-Strauss; The Anthropologist as Hero.* Edited by E. N. Hayes and Tanya Hayes. Cambridge, Mass.: M.I.T. Press.

CHERRY, COLIN. 1966. *On Human Communication.* Cambridge, Mass.: M.I.T. Press.

CHURCHILL, LINDSAY. 1971. "Ethnomethodology and Measurement." *Social Forces* 50, no. 2 (December).

CICOUREL, AARON. 1964. *Methodology and Measurement in Sociology.* New York: Free Press.

_____. 1968. *The Social Organization of Juvenile Justice.* New York: Wiley.

_____. 1974. *Cognitive Sociology.* New York: Free Press.

COMMAGER, HENRY STEELE. 1976. "Intelligence: The Constitution Betrayed." *New York Review of Books* 23, no. 15 (September 30).

COMMONS, JOHN R. 1961. *Institutional Economics.* Madison, Wis.: University of Wisconsin Press.

CONVERSE, PHILIP. 1964. "The Nature of Belief Systems in Mass Publics." *Ideology and Discontent.* Edited by David Apter. New York: Macmillan.

CORNFORD, JAMES. 1972. "The Political Theory of Scarcity." *Philosophy, Politics, and Society.* Edited by Peter Laslett, W. G. Runciman, and Quentin Skinner. Oxford: Blackwell

CROSSON, FREDERICK J., AND SAYRE, KENNETH, eds. 1967. *Philosophy and Cybernetics.* Notre Dame, Ind.: University of Notre Dame Press.

CURLE, ADAM. 1964. *World Campaign for Universal Literacy: Comment and Proposal.* Occasional Papers in Education and Development, Number One. Center for Studies in Education and Development, Cambridge, Mass.: Harvard University.

CURTIN, PHILIP. 1964. *The Image of Africa.* Madison, Wis.: University of Wisconsin Press.

DAEDELUS. 1959. 88, 4 (Fall). Special issue on quantity and quality.

DALY, ROBERT. 1970. "The Specters of Technicism." *Psychiatry* 33, no. 4 (November).

DEARDEN, R. F., ed. 1972. *Education and the Development of Reason.* London: Routledge & Kegan Paul.

DECHERT, CHARLES R. 1965. "Cybernetics and the Human Person." *International Philosophical Quarterly* 5, no. 1 (February).

DESMONDE, WILLIAM. 1971. "Gödel, Non-Deterministic Systems, and Hermetic Automata." *International Philosophical Quarterly* 11, no. 1 (March).

DEUTSCH, KARL. 1953. *Nationalism and Social Communication.* Cambridge, Mass.: M.I.T. Press.

————. 1966. *The Nerves of Government.* New York: Free Press.

DIAMOND, STANLEY, ed. 1964. *Primitive Views of the World.* New York: Columbia University Press.

————. 1974. *In Search of the Primitive: A Critique of Civilization.* New Brunswick, N.J.: Transaction Books.

DICKSON, PAUL. 1971. *Think Tanks.* New York: Atheneum.

DIJKSTERHUIS, E. J. 1961. *The Mechanization of the World Picture.* New York: Oxford.

DISCH, ROBERT, ed. 1973. *The Future of Literacy.* Englewood Cliffs, N.J.: Prentice-Hall.

DOUGLAS, JACK D. 1967. *The Social Meanings of Suicide.* Princeton, N.J.,: Princeton University Press.

————, ed. 1970a. *Deviance and Respectability.* New York: Basic Books.

————, ed. 1970b. *Understanding Everyday Life.* Chicago: Aldine.

DOUGLAS, MARY. 1970. *Natural Symbols.* New York: Pantheon.

DREEBEN, ROBERT. 1968. *On What Is Learned in School.* Reading, Mass.: Addison-Wesley.

DREYFUS, HUBERT. 1972. *What Computers Can't Do.* New York: Harper & Row.

DROR, YEHEZKEL. 1971a. *Ventures in Policy Sciences.* New York: American Elsevier.

————. 1971b. *Design for Policy Sciences.* New York: American Elsevier.

———. 1974. "Applied Social Science and Systems Analysis." *The Use and Abuse of Social Science.* Second edition. Edited by Irving L. Horowitz. New Brunswick, N.J.: Transaction Books.

DURKHEIM, EMILE. 1958. *Professional Ethics and Civic Morals.* New York: Free Press.

———. 1964. *The Division of Labor in Society.* New York: Free Press.

———. 1973. *Moral Education.* New York: Free Press.

EDEL, ABRAHAM. 1960–1961. "Science and Value: Some Reflections on Pepper's 'The Sources of Value.'" *Review of Metaphysics* 14, 1. (Sept.)

———. 1969. "Humanist Ethics and the Meaning of Human Dignity." *Moral Problems in Contemporary Society.* Edited by Paul Kurtz. Englewood Cliffs, N.J.: Prentice-Hall.

ELIADE, MIRCEA. 1959. *Cosmos and History.* New York: Harper & Row.

———. 1963. *Myth and Reality.* New York: Harper & Row.

ELKANA, YEHUDA. 1968. "The Emergence of the Energy Concept." Ph.D. dissertation, Department of History of Ideas, Brandeis University.

ELLUL, JACQUES. 1960. *The Theological Foundations of Law.* Garden City, N.Y.: Doubleday.

———. 1964. *The Technological Society.* New York: Random House Vintage.

———. 1965. *Propaganda.* New York: Knopf.

———. 1967a. *The Presence of the Kingdom.* New York: Seabury.

———. 1967b. *The Political Illusion.* New York: Knopf.

ERIKSON, KAI. 1966. *Wayward Puritans.* New York: Wiley.

ETZIONI, AMITAI. 1968. *The Active Society.* New York: Free Press.

EUBANK, EARLE E. 1932. *The Concepts of Sociology.* Boston: Heath.

FEINBERG, WALTER. 1975. "Educational Equality under Two Conflicting Models of Educational Development." *Theory and Society* 2, no. 2 (Summer).

FINE, SIDNEY. 1956. *Laissez Faire and the General Welfare State.* Ann Arbor, Mich.: University of Michigan Press.

FORD, JOHN J. 1966. "Soviet Cybernetics and International Development." *Social Impact of Cybernetics.* Edited by Charles Dechert. Notre Dame, Ind.: University of Notre Dame Press.

FOUCAULT, MICHEL.. 1975. "History, Discourse, and Discontinuity." *Psychological Man.* Edited by Robert Boyers. New York: Harper & Row.

FREIRE, PAULO. 1970a. *Pedagogy of the Oppressed.* New York: Herder and Herder.

———. 1970b. "Adult Literacy as Cultural Action for Freedom" and "Cultural Action and Conscientization." *Harvard Education Review* 40, nos. 2 and 3, respectively.

———. 1973. *Education for Critical Consciousness.* New York: Seabury.

FRENCH, PETER. 1974. *Conscientious Actions.* Cambridge, Mass.: Schenkman.

GALBRAITH, JOHN KENNETH. 1952. *American Captialism.* New York: New American Library.

GARFINKEL, HAROLD. 1967. *Studies in Ethnomethodology*. Englewood Cliffs, N.J.: Prentice-Hall.

GEERTZ, CLIFFORD. 1973. "The Cerebral Savage: On the Work of Claude Lévi-Strauss." *The Interpretation of Cultures*. New York: Basic Books.

GEIGER, THEODORE. 1967. *The Conflicted Relationship*. New York: McGraw-Hill.

GELLNER, ERNEST. 1974. *Legitimation of Belief*. Cambridge, Mass.: Cambridge University Press.

GEORGE, F. H. 1971. *Cybernetics*. London: English Universities' Press.

GOLLIN, GILLIAN. 1967. *Moravians in Two Worlds*. New York: Columbia University Press.

GOODMAN, PERCIVAL, AND GOODMAN, PAUL. 1960. *Communitas*. New York: Random House Vintage.

GOODY, JACK, ed. 1968. *Literacy in Traditional Societies*. Cambridge, Mass.: Cambridge University Press.

GOTESKY, RUBIN, AND LASZLO, ERVIN, eds. 1970. *Human Dignity*. New York: Gordon and Breach.

GOULDNER, ALVIN. 1965. *Enter Plato*. New York: Basic Books.

_____. 1970. *The Coming Crisis of Western Sociology*. New York: Basic Books.

_____. 1973. *For Sociology*. New York: Basic Books.

GRAHAM, LOREN. 1967. 'Cybernetics." *Science and Ideology in the Soviet Union*. Edited by George Fisher. New York: Atherton.

GRAMONT, SANCHE DE. 1970. "There Are No Superior Societies." *Claude Lévi-Strauss: The Anthropologist as Hero*. Edited by E. N. Hayes and Tanya Hayes. Cambridge, Mass.: M.I.T. Press.

GRAMPP, WILLIAM. 1960. *The Manchester School of Economics*. Stanford, Cal.: Stanford University Press.

GRAÑA, CÉSAR. 1964. *Modernity and its Discontents*. New York: Harper.

GRANT, GERALD AND RIESMAN, DAVID. 1978. *The Perpetual Dream: Reform and Experiment in the American College*. Chicago: University of Chicago Press.

_____, AND ASSOCIATES. *On Competence*. San Francisco: Jossey-Bass (forthcoming).

GREENBERG, GEORGE D. et. al. 1977. "Developing Public Policy Theory: Perspectives from Empirical Research." *American Political Science Review*. LXXI, 4, (December)

GROSS, BERTRAM. 1970. "Friendly Fascism." *Social Policy* 1, no. 4 (November–December).

GUNN, JAMES. 1961. *The Joy Makers*. New York: Bantam.

GUNN, J. A. W. 1969. *Politics and the Public Interest in the Seventeenth Century*. London: Routledge & Kegan Paul.

HAAS, ERNST. 1964. *Beyond the Nation-State: Functionalism and International Organization*. Stanford, Cal.: Stanford University Press.

_____. 1970. *Human Rights and International Action*. Stanford, Cal.: Stanford University Press.

HABERMAS, JÜRGEN. 1970. "Toward a Theory of Communicative Competence." *Recent Sociology, Number Two*. Edited by Hans Peter Dreitzel. New York: Macmillan.

_____. 1975. *Legitimation Crisis*. Boston: Beacon.

HAGE, JERALD. 1974. *Communication and Organizational Control*. New York: Wiley.

HAMPDEN-TURNER, CHARLES. 1970. *Radical Man*. Cambridge, Mass.: Schenkman.

HANKE, LEWIS. 1959. *Aristotle and the American Indians*. Bloomington, Ind.: Indiana University Press.

HARMON, DAVID. 1970. "Illiteracy: An Overview." *Harvard Education Review* 40, no. 2.

HARRISON, PAUL M. 1959. *Authority and Power in the Free Church Tradition*. Princeton, N.J.: Princeton University Press.

HART, H. L. A. 1955. "Are There Any Natural Rights?" *Philosophical Review*. 64. 2 (April).

Harvard Educational Review. 1970. "Illiteracy in America." 40, nos. 2 and 3.

HATT, HAROLD E. 1968. *Cybernetics and the Image of Man*. New York: Abingdon.

HAVELOCK, E. A. 1963. *Preface to Plato*. Cambridge, Mass.: Harvard University Press.

HAYDN, HIRAM. 1950. *The Counter-Renaissance*. New York: Harcourt, Brace.

HAYEK, F. A. 1948. *Individualism and Economic Order*. Chicago: University of Chicago Press.

_____. 1955. *The Counter-Revolution of Science*. New York: Free Press.

_____. 1967a. *Studies in Philosophy, Politics, and Economics*. Chicago: University of Chicago Press.

_____. 1967b. *The Road to Serfdom*. Chicago: University of Chicago Press/ Phoenix.

HAYWARD, J. E. S. 1960. "Solidarist Syndicalism: Durkheim and Duguit." *Sociological Review* 8, nos. 1 and 2.

HAZARD, PAUL. 1963. *The European Mind, 1680–1715*. New York: Meridian.

HEIDEGGER, MARTIN. 1950–1951. "The Age of the World View." *Measure* 2.

_____. 1967. *What Is a Thing?* Chicago: Regnery.

_____. 1968. *What Is Called Thinking?* New York: Harper & Row.

_____. 1969. *The Essence of Reasons*. Evanston, Il.: Northwestern University Press.

_____. 1976. "Only a God Can Save Us: *Der Spiegel*'s Interview with Martin Heidegger." *Philosophy Today* 20, no. 4 (Winter).

HEILBRONER, ROBERT. 1969. *The Limits of American Capitalism*. New York: Harper & Row.

_____. 1972. "Through the Marxian Maze." *New York Review of Books* 18, no. 4.

HELLER, ERICH. 1959. *The Disinherited Mind*. New York: Meridian.

_____. 1968. *The Artist's Journey into the Interior*. New York: Random House Vintage.

HELVEY, T. C. 1971. *The Age of Information*. Englewood Cliffs, N.J.: Educational Technology Publications.

HIEBERT, ERWIN. 1966. "The Uses and Abuses of Thermodynamics in Religion." *Daedelus* (Fall).

HIRST, P. H. 1972. "Liberal Education and the Nature of Knowledge." *Education and the Development of Reason*. Edited by R. F. Dearden. London: Routledge & Kegan Paul.

HOESS, RUDOLF. 1959. *Commandant of Auschwitz*. New York: Popular Library.

HOLLOWAY, JAMES. 1970. *Introducing Jacques Ellul*. Grand Rapids, Mich. Eerdmans.

HOLTMAN, ROBERT B. 1950. *Napoleonic Propaganda*. Baton Rouge, La.: Louisiana University Press.

HOOK, SIDNEY. 1943. *The Hero in History*. New York: Humanities Press.

HOOS, IDA R. 1972. *Systems Analysis in Public Policy*. Berkeley, Cal.: University of California Press.

HOPKINS, TERENCE. 1957. "Sociology and the Substantive View of the Economy." *Trade and Markets in the Early Empires*. Edited by Karl Polanyi. New York: Free Press and Falcon's Wing Press.

HOROWITZ, IRVING L., AND LIEBOWITZ, M. 1968. "Social Deviance and Political Marginality." *Social Problems*. 15 (Winter).

HORTON, JOHN. 1964. "The Dehumanization of Anomie and Alienation." *British Journal of Sociology* 15 (December).

HORTON, ROBIN, AND FINNEGAN, RUTH, eds. 1973. *Modes of Thought: Essays on Thinking in Western and Non-Western Societies*. London: Faber & Faber.

HUIZINGA, JOHANN. 1954. *The Waning of the Middle Ages*. Garden City, N.Y.: Doubleday.

Humanist, The. 1976. 36, no. 5 (September-October). Special issue on the Conlan amendment.

HUSSERL, EDMUND. 1962. *Ideas*. New York: Collier.

HYMES, DELL. 1973. "On the Origins and Foundations of Inequality among Speakers." *Daedelus* 102, no. 3 (Summer).

IGGERS, GEORGE. 1958. *The Cult of Authority*. The Hague: Nijhoff.

ILLICH, IVAN. 1970. *Celebration of Awareness*. Garden City, N.Y.: Doubleday.

_____. 1971. *Deschooling Society*. New York: Harper & Row.

_____. 1973. *Tools for Conviviality*. New York: Harper & Row.

JAEGER, WERNER. 1942–1944. *Paideia: The Ideals of Greek Culture*. Three volumes. Translated by Gilbert Highet. New York: Oxford.

JAMESON, FREDERICK. 1971. *Marxism and Form*. Princeton, N.J.: Princeton University Press.

_____. 1972. *The Prison-House of Language: A Critical Account of Structuralism and Russian Formalism*. Princeton, N.J.: Princeton University Press.

JAY, MARTIN. 1970. "The Megapolitics of Utopianism." *Dissent* 17, no. 4 (July-August).

JOHNSON, JAMES TURNER. 1975. *Ideology, Reason, and the Limitation of War.* Princeton, N.J.: Princeton University Press.

JONAS, HANS. 1958. *The Gnostic Religion.* Boston: Beacon.

_____. 1959. "The Practical Uses of Theory." *Social Research* 26, no. 2 (Summer).

KAHLER, ERICH. 1957. *The Tower and the Abyss.* New York: Viking.

_____. 1967. *Man the Measure.* New York: Meridian.

KAMENKA, EUGENE. 1962. *The Ethical Foundations of Marxism.* New York: Praeger.

KARABEL, JEROME, AND HALSEY, A. H. 1976. "The New Sociology of Education." *Theory and Society* 3, no. 4 (Winter).

KATZ, MICHAEL. 1971. *Class, Bureaucracy, and Schools.* New York: Praeger.

KENDRICK, THOMAS D. 1955. *The Lisbon Earthquake.* Philadelphia: Lippincott.

KIRK, KENNETH. 1927. *Conscience and Its Problems.* London: Longmans.

KLAUSNER, SAMUEL Z., ed. 1965. *The Quest for Self-Control.* New York: Free Press.

KLUCKHOHN, CLYDE. 1961. "The Study of Values." *Values in America.* Edited by Donald W. Barrett. Notre Dame, Ind.: Notre Dame Press.

KNUDSON, STEPHEN T. 1974. "The Education of the Amish Child." *California Law Review.* 62.

KOERNER, JAMES. 1973. "What Is Career Education?" Council for Basic Education, Occasional Papers no. 20 (February).

KOYRÉ, ALEXANDER. 1957. *From the Closed World to the Infinite Universe.* Baltimore: Johns Hopkins University Press.

KRADER, LAWRENCE. 1968. See entry under Marcel Mauss.

KRIEGER, LEONARD. 1965. *Politics of Discretion: Pufendorf and the Acceptance of Natural Law.* Chicago: University of Chicago Press.

KRIPPENDORFF, KLAUS. 1969a. "Theories and Analytical Constructs" and "Models and Messages: Three Prototypes." *The Analysis of Communication Content.* Edited by George Gerbner et al. New York: Wiley.

_____. 1969b. "Values, Modes, and Domains of Inquiry into Communication." *Journal of Communication* 19, no. 2 (June).

_____. 1969c. "On Generating Data in Communications Research." Ibid. 20, no. 3 (September).

_____. 1971. "Communication and the Genesis of Structure." *General Systems* 16.

KUHN, ALFRED. 1974. *The Logic of Social Systems.* San Francisco: Jossey-Bass.

KUHN, THOMAS S. 1957. *The Copernican Revolution.* New York: Random House Vintage.

KUNTZ, PAUL, ed. 1968. *The Concept of Order.* Seattle, Wash.: University of Washington Press.

LANDGREBE, LUDWIG. 1940–1941. "The World as a Phenomenological Problem." *Philosophy and Phenomenological Research* 1.

LANGE, FREDERICK ALBERT. 1925. *The History of Materialism.* London: Routledge & Kegan Paul.

LASZLO, ERVIN. 1971. "Human Dignity and the Promise of Technology." *Philosophy Forum* 9, no. 1/2 (June).

_____. 1973. *Introduction to Systems Philosophy*. New York: Harper & Row.

LEFEBVRE, HENRI. 1971. *Everyday Life in the Modern World*. New York: Harper & Row.

LENSKI, GERHARD. 1961. *The Religious Factor*. Garden City, N.Y.: Doubleday.

LÉVI-STRAUSS, CLAUDE. 1963. *Structural Anthropology*. New York: Basic Books.

_____. 1966. *The Savage Mind*. Chicago: University of Chicago Press.

LEWIS, C. S. 1965. *The Abolition of Man*. New York: Macmillan.

LILIENFELD, ROBERT. 1975. "Systems Theory as Ideology." *Social Research* 42, no. 4 (Winter).

LIN, NAN. 1973. *The Study of Human Communication*. Indianapolis: Bobbs-Merrill.

LOBKOWICZ, NICHOLAS. 1967. *Theory and Practice: History of an Idea from Aristotle to Marx*. Notre Dame, Ind.: University of Notre Dame Press.

LOEWITH, KARL. 1964. *From Hegel to Nietzsche: The Revolution in Nineteenth Century Thought*. New York: Holt, Rinehart & Winston.

LUKÁCS, GEORG. 1967. *History and Class Consciousness*. London: Merlin.

MACIVER, ROBERT, AND PAGE, CHARLES. 1959. *Society*. New York: Macmillan.

MACKAY, DONALD M. 1969. *Information, Mechanism, and Meaning*. Cambridge, Mass.: M.I.T. Press.

MACMURRAY, JOHN. 1936. *Interpreting the Universe*. London: Faber & Faber.

_____. 1957. *The Self as Agent*. London: Faber & Faber.

_____. 1961. *Persons in Relation*. London: Faber & Faber.

MACPHERSON, C. B. 1962. *The Political Theory of Possessive Individualism*. New York: Oxford.

_____. 1973. *Democratic Theory*. New York: Oxford.

MALIVER, BRUCE L. 1971. "Encounter Groupers Up Against the Wall." *New York Times Magazine* (January 3).

MANDELBAUM, MAURICE. 1955. *Phenomenology of Moral Experience*. New York: Free Press.

MANN, MICHAEL. 1970. "The Social Cohesion of Liberal Democracy." *American Sociological Review* 35, no. 3 (June).

MANNHEIM, KARL. 1936. *Ideology and Utopia*. New York: Harcourt, Brace.

_____. 1949. *Man and Society in an Age of Reconstruction*. New York: Harcourt, Brace.

MARCUSE, HERBERT. 1955. *Eros and Civilization*. Boston: Beacon.

_____. 1964. *One-Dimensional Man*. Boston: Beacon.

_____. 1968. *Negations: Essays in Critical Theory*. Boston: Beacon.

MARKS, STEPHEN R. 1977. "Multiple Roles and Role Strain: Some Notes on Human Energy, Time and Commitment." *American Sociological Review* 42, no. 6 (December).

MARUYAMA, MAGOROH. 1971. "The Second Cybernetics: Deviation Amplifying Mutual Causal Process." *Contemporary Sociological Theory*. Edited by Fred Katz. New York: Random House.

MARX, LEO. 1964. *The Machine in the Garden.* New York: Oxford.

MASSEY, J. L. 1968. "Information, Machines, and Men. *Philosophy and Cybernetics.* Edited by F. J. Crosson and K. M. Sayre. Notre Dame, Ind.: University of Notre Dame Press.

MATZA, DAVID. 1969. *Becoming Deviant.* Englewood Cliffs, N.J.: Prentice-Hall.

MAUSS, MARCEL. 1967. *The Gift.* New York: Norton.

_____. 1968. "A Category of the Human Spirit." *Psychoanalytic Review* 55, no. 3. See also the commentary on this reprint by Lawrence Krader in the same source.

MAYR, OTTO. 1970. *The Origins of Feedback Control.* Cambridge, Mass.: M.I.T. Press.

McCARTHY, T. A. 1973. "A Theory of Communicative Competence." *Philosophy of the Social Sciences* 3, no. 2 (June).

McCRACKEN, SAMUEL. 1970. "Quackery in the Classroom." *Commentary* 49 (June).

McCULLOCH, WARREN S. 1974. "Recollections of the Many Sources of Cybernetics." *Forum* 6, no. 2 (Summer).

McDERMOTT, JOHN. 1969. "Technology: The Opiate of the Intellectuals." *New York Review of Books* 13, no. 2.

McNEILL, JOHN. 1951. *A History of the Cure of Souls.* New York: Harper.

MEINECKE, FRIEDRICH. 1957. *Machiavellism: The Doctrine of Raison d'Etat and Its Place in Modern History.* New Haven, Conn.: Yale University Press.

MIKULAK, MAXIM W. 1966. "Cybernetics and Marxism-Leninism." *The Social Impact of Cybernetics.* Edited by Charles R. Dechert. Notre Dame, Ind.: University of Notre Dame Press.

MILGRAM, STANLEY. 1974. *Obedience to Authority.* New York: Harper & Row.

MILLS, C. WRIGHT. 1940. "Situated Actions and Vocabularies of Motive." *American Sociological Review* 5, no. 6 (December).

MIRANDOLA, PICO DELLA. 1956. *Oration on the Dignity of Man.* Translated by A. R. Caponigri. Chicago: Regnery.

MISES, LUDWIG VON. 1956. *The Anti-Capitalistic Mentality.* Princeton, N.J.: Van Nostrand.

_____. 1957. *Theory and History.* New Haven, Conn.: Yale University Press.

_____. 1960. *Epistemological Problems of Economics.* Princeton, N.J.: Van Nostrand.

_____. 1962a. *The Ultimate Foundations of Economic Science.* Princeton, N.J.: Van Nostrand.

_____. 1962b. *Bureaucracy.* New Haven, Conn.: Yale University Press.

MITCHAM, CARL, AND MACKEY, ROBERT, eds. 1972. *Philosophy and Technology.* New York: Free Press.

_____. 1973. "Bibliography of the Philosophy of Technology." *Technology and Culture* 14, no. 2, pt. 2 (April).

MITRANY, DAVID. 1966. *A Working Peace System.* Chicago: Quadrangle.

MUMFORD, LEWIS. 1962. *Technics and Civilization.* New York: Harcourt, Brace.

_____. 1966. *The Myth of the Machine.* New York: Harcourt, Brace.

_____. 1970. *The Pentagon of Power.* New York: Harcourt Brace Jovanovich.

NARVESON, JAN. 1967. *Morality and Utility.* Baltimore: Johns Hopkins University Press.

NELSON, BENJAMIN. 1964. "Actors, Directors, Roles, Cues, Meanings, Identities." *The Psychoanalytic Review* 51, no. 1 (Spring).

_____. 1965a. "Self-Images and Systems of Spiritual Direction in the History of European Civilization." *The Quest for Self-Control.* Edited by Samuel Z. Klausner. New York: Free Press.

_____. 1965b. "Probabilists, Anti-Probabilists, and the Quest for Certitude in the 16th and 17th Centuries." *Proceedings of the Xth International Congress for the History of Science*

_____. 1965c. "Dialogues Across the Centuries: Weber, Marx, Hegel, Luther." *The Origins of Modern Consciousness.* Edited by John Weiss. Detroit: Wayne State University Press.

_____. 1969. *The Idea of Usury.* Second edition. Chicago: University of Chicago Press.

_____, AND LUHMANN, NIKLAS. 1976. "A Conversation on Selected Theoretical Questions: Systems Theory and Comparative Civilizational Sociology." *Graduate Faculty Journal of Sociology of the New School* 1, no. 2 (Winter).

NISBET, ROBERT. 1966. *The Sociological Tradition.* New York: Basic Books.

_____. 1969. *Social Change and History.* New York: Oxford.

OAKESHOTT, MICHAEL. 1962. *Rationalism in Politics and Other Essays.* New York: Basic Books.

OLSON, DAVID E. 1974. *Media and Symbols.* Chicago: University of Chicago Press.

ONG, WALTER. 1969. "World as View and World as Event." *American Anthropologist* 71.

_____. 1971. *Rhetoric, Romance, and Technology.* Ithaca, N.Y.: Cornell University Press.

_____. 1974. *Ramus: Method and the Decay of Dialogue.* New York: Octagon.

OTTO, RUDOLF. 1967. *The Idea of the Holy.* New York: Oxford.

PALMER, RICHARD E. 1969. *Hermeneutics.* Evanston, Il.: Northwestern University Press.

PARKER, HARLEY. 1974. "The Beholder's Share and the Problem of Literacy." *Media and Symbols.* Edited by David Olson. Chicago: University of Chicago Press.

PARSONS, TALCOTT. 1966. *Societies.* Englewood Cliffs, N.J.: Prentice-Hall.

_____. 1967. *Sociological Theory and Modern Society.* New York: Free Press.

PATEMAN, CAROLE. 1970. *Participation and Democratic Theory.* Cambridge, Eng.: Cambridge University Press.

PATTEE, HOWARD H. ed. 1975. *Hierarchy Theory.* New York: Braziller.

PEPPER, STEPHEN. 1947. *A Digest of Purposive Values.* Berkeley, Cal.: University of California Press.

_____. 1958. *The Sources of Value.* Berkeley, Cal.: University of California Press.

_____. 1970. *World Hypotheses.* Berkeley, Cal.: University of California Press.

PETERS, R. S. 1972. "The Education of the Emotions." *Education and the Develop-*

ment of Reason. Edited by R. F. Dearden. London: Routledge & Kegan Paul.

PETTIT, PHILIP. 1975. *The Concept of Structuralism: A Critical Analysis.* Berkeley, Cal.: University of California Press.

PIAGET, JEAN. 1970. *Structuralism.* New York: Harper & Row.

POCOCK, J. G. A. 1968. "Time, Institutions, and Action." *Politics and Experience.* Edited by Preston King and B. C. Parekh. Cambridge, Eng.: Cambridge University Press.

―――. 1973. *Politics, Language, and Time.* New York: Atheneum.

―――. 1975. *The Machiavellian Moment.* Princeton, N.J.: Princeton University Press.

POLANYI, KARL. 1944. *The Great Transformation.* Boston: Beacon.

―――, ed. 1957. *Trade and Markets in the Early Empires.* New York: Free Press and Falcon's Wing Press.

PROSCH, HARRY. 1966. *The Genesis of Twentieth Century Philosophy.* Garden City, N.Y.: Doubleday.

PSATHAS, GEORGE. 1968. "Ethnomethods and Phenomenology." *Social Research* 35, no. 3 (Autumn).

REIMER, EVERETT. 1971. "An Essay on Alternatives in Education." *Interchange* 2, no. 1.

RICHARDSON, HERBERT W. 1967. *Toward an American Theology.* New York: Harper & Row.

RICHARDSON, WILLIAM J. 1963. *Heidegger: Through Phenomenology to Thought.* The Hague: Nijhoff.

RICHTER, MELVIN. 1964. *Politics of Conscience.* Cambridge, Mass.: Harvard University Press.

RICOEUR, PAUL. 1971. "The Model of the Text: Meaningful Action Considered as a Text." *Social Research* 38, no. 3 (Autumn).

ROBBINS, LIONEL. 1946. *An Essay on the Nature and Significance of Economic Science.* Second Edition, revised. New York: Macmillan.

ROSEN, LAWRENCE. 1971. "Language, History, and the Logic of Inquiry in Lévi-Strauss and Sartre." *History and Theory* 10, no. 3.

ROSEN, STANLEY. 1969. *Nihilism: A Philosophical Essay.* New Haven, Conn.: Yale University Press.

ROSZAK, THEODORE. 1969. *The Making of a Counter-Culture.* Garden City, N.Y.: Doubleday.

―――. 1972. *Where the Wasteland Ends.* Garden City, N.Y.: Doubleday.

ROTHMAN, DAVID J. 1971. *The Discovery of the Asylum.* Boston: Little, Brown.

RUDNICK, S. D. 1963. "From Created to Creator: Conceptions of Human Nature and Authority in 16th Century England." Ph.D. dissertation, Department of History of Ideas, Brandeis University.

RUNCIMAN, W. G. 1973. "What Is Structuralism?" *The Philosophy of Social Explanation.* Edited by Alan Ryan. New York: Oxford.

RUNES, DAGOBERT D. 1959. *Dictionary of Philosophy.* Ames, Iowa: Littlefield, Adams.

SAHLINS, MARSHALL. 1972. *Stone Age Economics*. Chicago: Aldine-Atherton.

SAINT-SIMON, HENRI DE. 1964. *Social Organization, the Science of Man and Other Writings*. Edited and translated by Felix Markham. New York: Harper.

SANDERS, RALPH. 1973. *The Politics of Defense Analysis*. New York: Dunellen.

SANTILLANA, GIORGIO DE. 1955. *The Crime of Galileo*. Chicago: University of Chicago Press.

SARTORI, GIOVANNI. 1965. *Democratic Theory*. New York: Praeger.

SARTRE, JEAN-PAUL. 1966. *Being and Nothingness*. New York: Washington Square Press.

SAVAS, E. S. 1970. "Cybernetics in City Hall." *Science* 169, no. 3395 (May 29).

SAYRE, KENNETH M. 1967. "Philosophy and Cybernetics." *Philosophy and Cybernetics*. Edited by F. J. Crosson and K. M. Sayre. Notre Dame, Ind.: University of Notre Dame Press.

_____. 1969. *Consciousness: A Philosophic Study of Minds and Machines*. New York: Random House.

_____. 1976. *Cybernetics and the Philosophy of Mind*. London: Routledge & Kegan Paul.

SCHEFF, THOMAS. 1966. *Being Mentally Ill*. Chicago: Aldine.

SCHICK, ALLEN. 1970. "The Cybernetic State." *trans-Action* 7, no. 4.

SCHNEIDER, LOUIS. 1971. "Dialectic in Sociology." *American Sociological Review* 36, no. 4 (August).

SCHUMACHER, E. F. 1973. *Small Is Beautiful*. New York: Harper & Row.

SCHUR, EDWIN M. 1971. *Labeling Deviant Behavior*. New York: Harper & Row.

SCHUTZ, ALFRED. 1963. "Common-Sense and Scientific Interpretations of Human Action." *Philosophy of the Social Sciences*. Edited by Maurice Natanson. New York: Random House.

_____. 1967. *The Phenomenology of the Social World*. Evanston, Il.: Northwestern University Press.

SCHUTZ, ALFRED, AND LUCKMANN THOMAS. 1973. *The Structures of the Life-World*. Evanston, Il.: Northwestern University Press.

SCHWARTZ, BENJAMIN. 1970. "The Religion of Politics." *Dissent* 17, no. 2 (March-April).

SCOTT, MARVIN B., AND LYMAN, STANFORD. 1968. "Accounts." *American Sociological Review* 33, no. 1 (February).

SCOTT, ROBERT A. AND DOUGLAS, JACK. 1972. *Perspectives on Deviance*. New York: Basic Books.

SCRIBNER, SYLVIA, AND COLE, MICHAEL. 1973. "Cognitive Consequences of Formal and Informal Education." *Science* 182, no. 4112 (November 9).

SEARLE, JOHN. 1969. *Speech Acts*. Cambridge, Eng.: Cambridge University Press.

SEWELL, JAMES PATRICK. 1966. *Functionalism and World Politics*. Princeton, N.J.: Princeton University Press.

SHILS, EDWARD. 1968. "The Concept of Consensus." *International Encyclopedia of the Social Sciences*. Volume 3. Edited by David Sills. New York: Free Press and Macmillan.

SJOBERG, GIDEON. 1959. "Operationism and Social Research." *Symposium on Sociological Theory.* Edited by Llewellyn Gross. New York: Harper.

SKINNER, B. F. 1948. *Walden Two.* New York: Macmillan.

SMITH, PERRY McCOY. 1970. *The Air Force Plans for Peace, 1943–1945.* Baltimore: Johns Hopkins University Press.

SNOOK, I. A. 1972. *Concepts of Indoctrination.* London: Routledge & Kegan Paul.

SOLOMON, ROBERT C. 1976. *The Passions.* Garden City, N.Y.: Doubleday.

SPENCER, ROBERT F., ed. 1969. *Forms of Symbolic Action: Annual Proceedings of the American Ethnological Society.* Seattle, Wash.: University of Washington Press.

SPIEGELBERG, HERBERT. 1971. "Human Dignity: A Challenge to Contemporary Philosophy." *Philosophy Forum* 9, no. 1/2 (March).

———. 1973. "The Right to Say We." *Phenomenological Sociology.* Edited by George Psathas. New York: Wiley.

SPRAGENS, THOMAS. 1973. *The Politics of Motion.* Lexington, Ky.: University of Kentucky Press.

STANLEY, MANFRED. 1967. "Social Development as a Normative Concept." *Journal of Developing Areas* 1, no. 3.

———. 1969. "Jehovah in the City of Mammon." *Urbanism, Urbanization, and Change.* Edited by E. H. Mizruchi and Paul Meadows. Reading, Mass.: Addison-Wesley.

STEINBRUNER, JOHN D. 1974. *The Cybernetic Theory of Decision.* Princeton, N.J.: Princeton University Press.

STEINER, GEORGE. 1970. *Language and Silence.* New York: Atheneum.

———. 1973. "After the Book?" *The Future of Literacy.* Edited by Robert Disch. Englewood Cliffs, N.J.: Prentice-Hall.

STERN, FRITZ. 1965. *The Politics of Cultural Despair.* Garden City, N.Y.: Doubleday.

STINCHCOMBE, ARTHUR. 1968. *Constructing Social Theories.* New York: Harcourt, Brace.

STOKES, RANDALL, AND HEWITT, JOHN P. 1976. "Aligning Actions." *American Sociological Review* 41, no. 5 (October).

SYPHER, WYLIE. 1968. *Literature and Technology: The Alien Vision.* New York: Random House.

SZTOMPKA, PIOTR. 1974. *System and Function.* New York: Academic Press.

TALMON, J. L. 1960. *Political Messianism: The Romantic Phase.* New York: Praeger.

TAYLOR, CHARLES. 1971. "Interpretation and the Sciences of Man." *Review of Metaphysics* 25, no. 1 (September).

TAYLOR, RICHARD. 1973. *Freedom, Anarchy, and the Law.* Englewood Cliffs, N.J.: Prentice-Hall.

THEOBALD, ROBERT. 1961. *The Challenge of Abundance.* New York: New American Library.

———. 1965. *Free Men and Free Markets.* Garden City, N.Y.: Doubleday.

THIELICKE, HELMUT. 1969. *Nihilism: Its Origin and Nature*. New York: Schocken.

TOENNIES, FERDINAND. 1957. *Community and Society*. Translated by C. P. Loomis. East Lansing, Mich.: Michigan State University Press.

TYLOR, STEPHEN A. 1969. *Cognitive Anthropology*. New York: Holt, Rinehart & Winston.

UNDERWOOD, KENNETH. 1957. *Protestant and Catholic*. Boston: Beacon.

VOEGELIN, ERIC. 1952. *The New Science of Politics*. Chicago: University of Chicago Press.

WALSH, VIVIAN CHARLES. 1961. *Scarcity and Evil*. Englewood Cliffs, N.J.: Prentice-Hall.

WARNER, W. LLOYD. 1961. *The Family of God*. New Haven, Conn.: Yale University Press.

WARREN, ROLAND L. 1963. *The Community in America*. Chicago: Rand McNally.

WATTS, ALAN. 1964. *Beyond Theology*. New York: Pantheon.

WEBER, MAX. 1958. "Politics as a Vocation." *From Max Weber*. Edited by Hans Gerth and C. Wright Mills. New York: Oxford.

WEBSTER, MAUREEN. 1971. "Educational Planning in Transition." Ph.D. dissertation, Syracuse University School of Education.

WEBSTER, MAUREEN; CLASBY, MIRIAM; AND WHITE, NAOMI. 1973. *Laws, Tests, and Schooling*. Educational Policy Research Report RR-11 (October). Published by Educational Policy Research Center, Syracuse Research Corporation.

WEISSKOPF, WALTER A. 1955. *The Psychology of Economics*. Chicago: University of Chicago Press.

_____. 1971. *Alienation and Economics*. New York: Dutton.

WEIZENBAUM, JOSEPH. 1976. *Computer Power and Human Reason*. San Francisco: Freeman.

WHITE, HAYDEN. 1973. "Foucault Decoded: Notes from Underground." *History and Theory* 12, no. 1.

WHITE, LESLIE. 1949. *The Science of Culture*. New York: Grove Press.

WHYTE, L. L., WILSON, A. G., AND WILSON, D., eds. 1969. *Hierarchical Structures*. New York: American Elsevier.

WIEBE, ROBERT H. 1967. *The Search for Order, 1877–1920*. New York: Hill and Wang.

WIENER, NORBERT. 1954. *The Human Use of Human Beings*. Garden City, N.Y.: Doubleday.

_____. 1965. *Cybernetics*. Cambridge, Mass.: M.I.T. Press.

WILENSKI, HAROLD. 1964. "Mass Society and Mass Culture: Inderdependence or Independence?" *American Sociological Review* 29.

WILKINSON, RUPERT. 1964. *Gentlemanly Power*. New York: Oxford.

WILLEY, BASIL. 1953. *The Seventeenth-Century Background*. Garden City, N.Y.: Doubleday.

WINNER, LANGDON. 1977. *Autonomous Technology*. Cambridge, Mass.: M.I.T. Press.

WINTHROP, HENRY. 1968. *Ventures in Social Interpretation.* New York: Appleton-Century-Crofts.

WISCONSIN V. YODER. 1972. 406 U.S. 205.

WOLFF, ROBERT PAUL. 1970. *In Defense of Anarchism.* New York: Harper & Row.

WOLIN, SHELDON. 1960. *Politics and Vision.* Boston: Little, Brown.

WOOLF, HARRY ed. 1961. *Quantification.* Indianapolis: Bobbs-Merrill.

WRONG, DENNIS. 1961. "The Oversocialized Conception of Man in Modern Sociology." *American Sociological Review* 26, no. 2 (April).

_____. 1966. "The Idea of Community: A Critique." *Dissent* 13.

ZAHN, GORDON. 1964. *In Solitary Witness.* New York: Holt, Rinehart & Winston.

ZIJDERVELD, ANTON. 1970. *The Abstract Society.* Garden City, N.Y.: Doubleday.

ZWEIG, PAUL. 1968. *The Heresy of Self-Love.* New York: Basic Books.

Index

273